THE WOODWORM PROBLEM

THE RENTOKIL LIBRARY

Adult and larva of *Anobium punctatum* much enlarged. Although called the Common furniture beetle, this species is by far the most important pest of softwood in buildings in Britain.

THE RENTOKIL LIBRARY

THE WOODWORM PROBLEM

NORMAN E. HICKIN

Scientific Consultant
Rentokil Limited

RENTOKIL LIMITED
EAST GRINSTEAD

First published 1963
Second edition 1972
by
Hutchinson & Co (Publishers) *Ltd.*

Third (revised) edition 1981
published by RENTOKIL LTD.
Felcourt, East Grinstead,
W. Sussex RH19 2JY

ISBN 0 906564 03 4

Filmset, printed and bound in Great Britain by
Hazell Watson & Viney Ltd, Aylesbury, Bucks

Contents

Preface

This third edition of *The Woodworm Problem* retains most of the text from the second edition. Only minor changes have been made to the chapters on insects and fungi since their biology and habits do not alter. Statistics concerning the various wood-boring species have, however, been updated. The wood-boring weevil species have been given a chapter on their own owing to their increased importance and note has been taken of the changed status of the Wharf borer. Developments in control have been described and most importantly, a section on Health and Safety added.

The changes noted above have been made by Robin Edwards, the editor of the Rentokil Library, well-known for the painstaking way in which he conducts this work and for which the author is profoundly grateful.

Introduction

Wood is a very important material in our national economy. It is the most versatile substance used by man and has played an exceptional part in the evolution of his civilization. Perhaps the dawn of civilization was when fire was first controlled – maybe at the edge of a smouldering forest fire. Certainly, burning for warmth or cooking is still one of its most important uses today. But since the earliest times man has fashioned wood for almost every conceivable purpose – weapons and tools, boats and wheeled carts, musical instruments and buildings.

Today an immense amount of wood is in use in Britain. A very high proportion of the 22 000 000 houses are of partly timber-framed construction and in addition a high proportion of the domestic furniture in use today is made of wood. These two basic uses of wood are the main subject of this present account. It is not generally realized that the annual value of wood imported into Britain is over £1000 million – say £3 million each day. Home-grown timber is about 10% of this figure, but due to the maturation of forests it is likely to increase substantially within the next ten years, but most of our timber arrives at our ports after a long and costly journey and, moreover, due to several other causes, timber is becoming increasingly expensive. It seems a logical conclusion, therefore, to look to the conservation of timber to make it last longer. This is of great importance because, as will be explained in a following chapter, it seems that the more wood we use, the shorter time it is likely to last, *unless we do something about it.*

1   What causes wood to decay

Before discussing the special problem of woodworm we should first of all consider the factors and agencies causing the destruction or decay of wood and thus see where wood-boring insects enter the general picture. Wood is an organic substance – the woody tissue elaborated by certain plants – and, in common with all other substances which are the product of living processes, can be consumed by other living organisms which cause it to deteriorate. It should be thoroughly understood that 'timber does not deteriorate through age alone'. The factors or agencies involved may be vegetable, animal, mineral or physical and can be summarized as follows:

Factors and agencies causing destruction or decay of wood
1. *Attack by fungi:* Fungi are plants, members of the vegetable kingdom. They differ, however, from all other plants by not possessing the green colouring matter chlorophyll, with the aid of which energy from sunlight is absorbed to convert the gas, carbon dioxide, and water to sugar and starch. Fungi then must be either parasitic – deriving their nourishment from living organic tissue such as the fungi often present on living trees, or saprophytic – deriving their nourishment from organic tissue already dead or decaying. The Dry rot fungus – *Serpula lacrymans* – is an example of this type.

The fungal hyphae – the minute delicate threads which rot the wood by feeding on a constituent of it (the cellulose) – can only be seen after sectioning, staining and mounting under the high-power objective of the microscope, but certain 'gross' features of fungal attack are always present when such decay has proceeded for some time. The wood becomes light brownish in colour and cracks, either transversely (across the grain) (Fig. 1), or longitudinally. If the wood is painted or otherwise decorated, then severe buckling usually takes place. Considerable weight is lost and a 'sporophore' or fruiting body which produces the spores may ultimately develop.

The more obvious wood-rotting fungi, such as the Dry rot fungus *Serpula lacrymans*, belong to a group known as the BASIDIOMYCETES, but in recent years attention has been drawn to a group of primitive fungi

11

Fig. 1. Wood showing attack by the Dry rot fungus – *Serpula lacrymans*.
Note cracking into cubes and patches of white mycelium.

exemplified by the species *Chaetomium globosum* which causes a superficial rot on the surface of wood known as 'Soft rot'. This type of rot is
known to favour the establishment of several groups of wood-boring
animals. The presence of other types of fungal decay is essential for the
attack by several species of wood-boring beetles, described in greater
detail in later chapters.

Fungal decay is extremely important in wood used out-of-doors
whereas indoors it is much less important than woodworm attack. Fungal
decay requires a moisture content in the wood greater than that normally
found in buildings except when there is accidental leakage, faulty building maintenance or in old buildings lacking a damp-proof course. No
comprehensive account of fungal decay is given in this book as such
accounts already exist. The explanation of fungal decay which is given
here will be sufficient for the reader to understand this phenomenon.
References to books giving fuller accounts of fungal decay are given in
the list of references at the end of this book.

2. *Attack by wood-boring animals:* This section includes insects such as
beetles (COLEOPTERA) – the beetles whose larvae are known as woodworm are placed here, wood-boring wasps (HYMENOPTERA), termites
(ISOPTERA), etc., CRUSTACEA such as *Limnoria*, known as the 'Gribble',
MOLLUSCA such as *Teredo*, known as 'Shipworm' (Fig. 2). The Gribble
and the Shipworm attack only wood floating or submerged in the sea
and are known as the marine borers. They were, and still are to a much
more limited extent, grave hazards to wooden-hulled ships.

Fig. 2. Sawn timber showing attack by the shipworm, *Teredo*. Note white calcareous lining to the tunnels. This is a valuable character for indentifying this type of attack.

3. *Mechanical wear:* The effects of abrasion (Fig. 3) can be observed on well-used unprotected wooden stairs. Often sliding drawers in furniture abrade badly, throwing off a fine dust often confused with the bore-dust of woodworm.

4. *Chemical degradation or decomposition of wood:* Many chemicals have the effect of destroying or weakening wood to a serious extent. Acids and alkalis have this effect (Fig. 4). Often wood in contact with acids for a prolonged period has a woolly appearance, due to the separation of the wood fibres. One phenomenon rightly belonging to this category is the interaction of certain timbers on iron nails. The complete

Fig. 3. Mechanical wear is another factor causing decay of wood. Probably mainly caused by abrasion, innumerable tiny cuts causing the timber to wear away. The illustration shows a wooden 'pebble'. A block of wood has had all the corners worn off by the action of waves washing it up and down the sea shore.

Fig. 4. Wood decayed by chemical action. An alkali, sodium hydroxide solution, has caused the destruction of this wood.

Fig. 5. Decay caused by physical factors. Sample of timber showing prolonged exposure to moderate heat giving appearance not unlike that due to some types of fungal attack. Temperatures well below that of boiling water if sustained for a long period often cause such decay. Warm water decay is sometimes associated with fungal attack of a special kind.

Fig. 6. Bacteria cause extremely little decay of wood but we illustrate one of the rare instances – a piece of pine piling put into position under the City of Rotterdam in the sixteenth century and blown out during the bombing in 1940 from a position eighteen feet deep where it lay in wet mud. The bacteria causing this decay are of the anaerobic type – that is, they live and decay the timber in the absence of oxygen.

rusting of iron nails in oak and other timbers with a simultaneous cracking of the wood is well known.

5. *Heat:* Whilst prolonged low temperature has little or no effect on wood (provided the moisture content of the wood is not unduly high), prolonged heating even at temperatures below that of boiling water causes breakdown of the wood – often simulating fungal attack to a remarkable degree (Fig. 5).

6. *Bacteria:* Bacteria play an almost negligible role in decay of wood, yet their presence has been shown in certain situations. The presence of anaerobic (living in the absence of oxygen) bacteria decaying the outer skin of softwood piles lying many feet in wet silt under the City of Rotterdam has been demonstrated. The bacteria-decayed wood is similar to fungal attack. It is discoloured to a greyish-brown and there are innumerable fine transverse cracks (Fig. 6).

In each category of the above summary, some timbers withstand the agencies of destruction to a greater extent than others. In other words some timbers are more *durable* than others, with particular factors in mind. It does not follow that a species of timber is durable against all possible factors.

What is woodworm?

Woodworm is the name given to the larval or immature stage of certain wood-boring beetles found in buildings. Strictly speaking, then, they are not worms but are analogous to the grub or caterpillar stage of a moth or butterfly. Until the 1920s the term 'woodworm' meant only the larvae of the Common furniture beetle (*Anobium punctatum*) or Death watch (*Xestobium rufovillosum*) as the various *Lyctus* species were only then becoming common after introduction from countries abroad. It is interesting to note that in New Zealand the Common furniture beetle

(*Anobium punctatum*), which is exactly the same species as found in Britain, is known by the much more applicable name of 'House-borer'. This is in view of the fact that for every furniture beetle larva tunnelling in furniture there must be thousands tunnelling in the joists, rafters and floors of houses.

What are the gross features of a woodworm infestation; that is, what are the naked eye signs and how may it be distinguished from fungal attack?

Normally, the presence of flight holes, usually circular, is the first sign of woodworm infestation. The adult or beetle stage, if in sufficient numbers, may also be observed, but it is amazing how often beetles may be present in hundreds, yet the householder is unaware of their existence, or even if they are observed, their connection with woodworm is not inferred. The presence of 'dust' or wood powder is often associated with the flight or exit holes, very much more so in the various species of *Lyctus* than Furniture beetle and other species. This is brought about by new boring intercommunicating with an old tunnel leading to a flight hole. It is important to remember that the powder is kicked by a larva out of a flight hole through which it will *not* itself emerge. Sometimes powdering is caused by predacious insects which are feeding on the woodworms.

An insect is an animal distinguished from all other animals by having six jointed legs in the adult stage, with a segmented body. During its life cycle it undergoes a complete metamorphosis (that is, the change from grub or larva through pupa or chrysalis to adult), or the metamorphosis may not be complete – the young individuals looking very much like the adults except in the possession of wings. Wood-boring beetles are examples of the former type with a complete metamorphosis (Fig. 7). Regular circular tunnels in the wood (Fig. 8), flight holes on the exterior surface (Fig. 9), the associated grub-like larvae inside the wood and the resulting beetles outside the wood are all characteristic of woodworm damage, easily distinguished from fungal decay described previously.

Which species does most damage?

What is the relative importance of the various species of wood-boring beetles? If we assess the damage caused by wood-boring beetles in the United Kingdom – damage caused to all fabricated woodwork such as furniture, structural timber, floorings and indeed all the things that man has fashioned out of wood – and then assess the proportion of the whole caused by each species of wood-boring beetle, what sort of result would we get? A truly scientific approach to this question at the present time would be very difficult. What units would we employ? The number or population of various insects would obviously not be an accurate guide. It would take several hundred furniture beetle larvae to do the same damage as one House longhorn larva. Again there are quite local areas where one species predominates – such as an area of Surrey where

Fig. 7. The life-cycle of Common furniture beetle: *Anobium punctatum.* This shows an insect with complete metamorphosis, i.e. going through a larval and pupal stage, the larval stage being very different in outward appearance from that of the adult beetle. The life-cycle is passed through in about three years but it is the 'worm' or grub stage which is responsible for most of the damage, although the adult beetle biting its way out is responsible for the little holes, the only visible damage.

Fig. 8. Deal floorboard with the surface planed to show the tunnelling inside by *Anobium punctatum.* Damage equal to this can seriously impair the load-carrying properties.

Fig. 9. An attack of Common furniture beetle out-of-doors in the branch scar of an old apple tree.

House longhorn occurs – but when viewed against the country as a whole, a very different perspective is obtained.

Another difficulty in assessing both the relative importance and the actual incidence of woodworm attack, if the word 'attack' or 'infestation' is used, is to say what constitutes an infestation – one flight hole or a thousand? Obviously agreement or a rule regarding this must be obtained by the authorities concerned.

The relative occurrence of the various wood-boring beetles is now known accurately (actually previous estimated figures were found to be fairly accurate) and in addition variations in the distribution of the different species in the various parts of the United Kingdom are known. The figures, published by the author in 1967, show in a graphic way the various proportions of wood-boring pests according to the statistics gathered from Rentokil surveys. During a 5½-year period from 1960 to 1965, the timber in 142 000 buildings was examined and infestations by wood-boring insects and wood-rotting fungi reported upon. It should be stated that the owners of all these buildings had requested the company to make the survey so that it must be realized that the 'sample' of buildings was not a 'random' sample, as many of the owners must have at least known that the condition of the timber in the buildings was suspect. As far, however, as the comparative incidence of wood-boring insects is concerned, this does not matter.

It is of interest to note how this sample is made up in respect of types of buildings, which were as follows: 84·8% private dwelling houses; 6·5% commercial premises; 4·5% churches and historic buildings; 1·0% licensed premises and 3·2% farms. The age distribution of the buildings in the sample is also of interest and was as follows: pre-1919, 53·4%; 1920–39, 40·5% and 1940–62, 5·2%.

A further statistical survey was made in 1972 when the incidence of the major pest insects had changed slightly, the wood-boring weevils being found more frequently. The figures for this one-year survey are shown in brackets after the main survey percentages in the following résumé.

Common furniture beetle *Anobium punctatum*
The important result of the survey, which surprises many people, is the great preponderance of damage caused by the Common furniture beetle: 74·9% (83·4%) of the buildings surveyed showed sufficient woodworm damage by this insect as to merit control by a servicing company. Throughout the United Kingdom the comparative importance of *Anobium punctatum* is uniformly high but is highest in the English southern counties, the figure being in excess of 80%. It was found to be lowest in the north-western counties at 59·6%, whilst in Scotland it was 70·9%. Quite independently of these results it was found in 1962 by the officers of a local authority that 85% of pre-1939 houses in a council estate in Shoreham, Sussex, had 'Woodworm'.

Wood-boring weevils *Euophryum* species and *Pentarthrum huttoni*
Attack by these species was found in 5·2% (9·5%) of all buildings and was particularly abundant in London, occurring in 15·6% of buildings. It occurred only rarely in Scotland. Exact identifications were not made as it is difficult in the field, but the former species is thought to dominate. Note the marked increase in abundance of these weevils in the two surveys.

Death watch beetle *Xestobium rufovillosum*
The number of infestations attributed to Death watch amounted to 4·7% (4·7%) of all the buildings surveyed. It was not found in Scotland nor in Northern Ireland. This was to be expected, as this insect is extremely rare in those countries. The relative abundance of this beetle remained fairly constant during the 1960s and 1970s.

Powder-post beetles *Lyctus* species
Lyctus beetle attack was found only in 0·9% (0·4%) of the buildings examined. At one time it was thought to be a much more serious problem. In the north of England and Scotland the figure was only about half that given.

House longhorn beetle *Hylotrupes bajulus*
The House longhorn prefers the micro-climate of the roof spaces of houses in which to carry out its life cycle, and thus often escapes detection. In the 5½-year survey, only 533 infestations were found, 456 of

which were in Surrey and London, being 0·3% of all buildings examined in those areas. In the country as a whole it was found in 0·4% of all buildings. The 1972 survey showed the numbers reduced to only 0·1% of all buildings.

Other beetles
A large number of beetles share the responsibility for the last small fraction of damage. A few of the more common ones are described later.

Incidence of woodworm
The 'true' or 'actual' incidence of woodworm is, of course, something quite different from the comparative occurrence of the different species. We are now considering the total number of infested buildings in the United Kingdom. A survey of 22 000 000 buildings would be a colossal task to perform but, as carried out in many other types of survey, a cross-section or sample can be examined.

There are many problems in selecting buildings for the sample survey in making sure that the proportions of the various types of buildings in the sample are fully representative in all respects that are likely to influence, one way or another, the incidence of the various wood-boring insects. A thorough study must be made of all the factors, biological and social, before an authoritative survey can be made. On the other hand there are certain indications which point to such a substantial proportion of Furniture beetle incidence that mention should be made of it here. During part of 1961, 694 timber surveys were carried out in respect of applications for insurance against woodworm. There is reason to believe that the owners of a large proportion of these properties were under the impression that woodworm did not occur in their homes. If is of interest to note the age groupings of this sample of buildings. 26·2% were built before 1919, 47·9% were built between 1920 and 1939 and 13·9% built between 1940 and 1962. Of the 694 buildings, 415 or 59·8% contained an active infestation of *Anobium punctatum*. Of the 166 buildings surveyed that were built between 1940 and 1962 – what we can term *new* houses – 36, or 21·7% were infested with *Anobium punctatum*. At the present stage of our study of the nature and extent of the statistics available, it seems certain to the writer that at least half the buildings in the United Kingdom contain an active infestation by the wood-boring beetle *Anobium punctatum*, and of the houses built since the end of the last war it is probable that something like one in five is infested.

Where does it come from?
Where does woodworm come from? Has woodworm a habitat outside man's economy? How does woodworm enter a house in the first place? Can it fly? Here are questions which everyone asks when woodworm of one species or another enters his own home.

Now Furniture beetles, Powder post beetles and Death watch are found quite commonly out-of-doors. Indeed Death watch is much more frequently met with outside that within. Many hardwood trees – oak, elm, ash and especially willow – which are dead, perhaps struck by lightning, or with some dead part such as a branch or wound where a branch has been lopped off, will harbour some Death watch. Old pollard willows are perhaps the most fruitful souce of these insects and the dead wood in the crown of the tree always seems to support a good colony. Death watch is found in such situations where fungal decay is evident, whereas Furniture beetle, on the other hand, is seen to occur where for some reason the wood has resisted the onset of fungal attack, possibly remaining fairly dry by sheltered position or good drainage. Old branch scars on trunks of trees or areas of exposed sapwood are almost always populated by Common furniture beetle (Fig. 9). The freedom from fungal decay in such situations where this insect is established leads the writer to conjecture whether some fungicidal material is secreted into the timber by the developing larvae.

Lyctus, too, occurs in timber out-of-doors. It attacks, however, only the sapwood of certain hardwoods such as oak, which have been newly felled, ringed by animals, uprooted by storm, or where for some other reason the sap flow has ceased. Only for a few years does *Lyctus* maintain itself on such timber – only while the starch content in the sapwood is high. Only perfectly sound timber is attacked. *Lyctus* is never found in timber already attacked by fungus. It is most commonly found in timber yards.

We have thus dealt with the question of where woodworm occurs outside our homes – at least with the three most common species. Now we must see how they get into our homes. With Common furniture beetle and Powder post the explanation is simple. The adult beetle possesses a good pair of wings tucked underneath the wing-cases (the elytra) and both species fly readily (Fig. 10). Common furniture beetle in flight may often be mistaken for a house fly, even by the experienced entomologist, so closely does it resemble the latter, even to flying round central light-coloured objects in a room, such as electric light bulbs.

Fig. 10. Common furniture beetle showing position of open wing-cases and wings ready for flight.

Beetles that have made a flight can nearly always be identified as they are not usually able to fold the tips of their wings under the wing-cases once they have flown. Furniture beetle is often met with out-of-doors and beetles may settle on a person's clothing and be brought indoors. It can, of course, fly through the open window in either direction. As far as flight is concerned *Lyctus* is very similar in habit to Common furniture beetle. It does seem, however that *Lyctus* is attracted to light to a greater degree than Common furniture beetle and the adult beetle may often be seen in the flight season on the window panes where an infestation occurs in a room. It is not known how far these two species can fly but their flight is quite strong enough for them to be carried considerable distances if caught by the wind.

Death watch beetle very seldom flies, except at temperatures in excess of about 21–22°C and it seems certain that this insect is introduced into a building with the original timbers. The latter were invariably of large section and cut from over mature oak and chestnut trees already infested with a wood-rotting fungus and containing the Death watch larvae.

Many wood-borers, particularly Powder post beetles and Longhorn beetles, are being introduced mostly from overseas into our homes in furniture, wooden ornaments and packing cases constructed from infested timber. Sometimes the infestation remains undetected for several years until finally the affected article collapses.

Notes on the scientific names of insects

The reader will have observed in going through this chapter that often when the common (English) name of an insect or plant is given it is followed by the scientific, or latinized words, always printed in italics. You will want to know the reason for this, which is as follows: the common names of our animals and plants differ in various localities and in addition would be quite unknown to foreign biologists. It would be quite unusual for the common names of insects to be found in English dictionaries! Biologists have therefore evolved a system of naming animals and plants which has been universally adopted and is known as the Linnaean system (named after is founder, Linnaeus – 1707–78). These rules of naming are rather complicated but very exact, and although anomalies do arise from time to time, taken generally the system works very well. Expressed very simply it is as follows.

A natural classification is the basis of the system. That is, all the animals and plants are·arranged in the way that is thought that they evolved and names are given to the important groups that have characters in common. For instance, the 'class' INSECTA is divided into a number of 'orders', examples being COLEOPTERA (beetles), LEPIDOPTERA (butterflies and moths) and DIPTERA (two-winged flies). The orders are again divided into families and the families sub-divided into the smallest groups, which are called genera. Each one of these latter is called a genus. The individual species are grouped into the genera. Each genus is given a name consisting of a single word and each sort of individual likewise. The name given to a genus is never duplicated, but this may be done in the case of the species name. Thus, the combination of the name of the genus and the name of the individual species means one particular animal, and furthermore its relationship with a group of animals will be seen at once. The generic name is always commenced with a capital letter and the individual or trivial name as it is sometimes called, is always commenced with a small letter. Both words are always underlined in manuscript and typescript to show that, if printed, the whole name of the insect would appear in italics. More correctly the name of the author (who first described the species and named it) should be given afterwards, either in full or in an abbreviated form, but this rule need not be adopted on every occasion.

2 Common furniture beetle

The great importance of the Common furniture beetle has already been stated – it is clearly the most serious wood-boring insect in the United Kingdom. Here is given a description of the biology of this beetle, its feeding habits and the timber infested.

Common furniture beetle is the colloquial name given to the insect which de Geer called *Anobium punctatum* in 1774. (*Anobium punctatum* de Geer is referred to in literature as *Anobium striatum*; this was the name given to the insect by Olivier in 1790; and also as *Anobium domesticum* given by Geoffroy in 1785. In accordance with the law of priority these latter names are now not used.)

In those days and indeed until relatively recent years *Anobium punctatum* was a beetle associated with old houses and old furniture. For instance, Stephens in 1839 said that this beetle was 'abundant in old houses throughout the country' whilst in 1931 Joy, in his *Practical Handbook of British Beetles*, said 'the larvae attack furniture which is said to be "worm-eaten" '. Today the situation is very different. Only some fifty

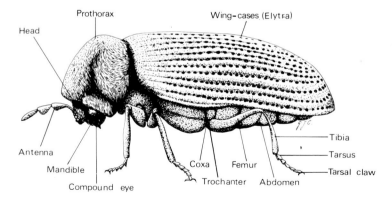

Fig. 11. Side view of Common furniture beetle showing the hood and the longitudinal rows of 'punctures' on the wing-cases.

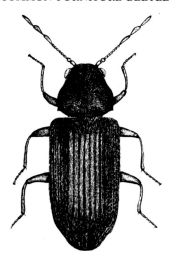

Fig. 12. Common furniture beetle – view from the top showing how the head is hidden by the prothorax. It is 2·5–5 mm in length and dark chocolate-brown in colour.

years since Joy wrote this, as has already been stated, half the houses in the British Isles are probably attacked to some degree or another by this beetle. The writer's theories to account for this increase are given later, after the section dealing with the life-cycle and the types of woods attacked. They will then be better understood.

The Common furniture beetle belongs to the family ANOBIIDAE which contains about twenty-two different species found in Britain. This total includes the Death watch beetle and several other species of economic importance. Among these is *Stegobium paniceum*, often known as the Biscuit beetle, which is a common pest of stored foods, soup powders, etc. (It is often mistaken for the Furniture beetle, as it has a superficial resemblance to this species.)

The adult beetle varies in size from 2·5 to 5 mm in length and it is usually chocolate-brown in colour, while lighter and darker individuals are often encountered. All are covered with a fine yellowish down which does not, however, hide the longitudinal rows of 'punctures' or small pits on the 'elytra' or wing-cases. The most characteristic feature, however, of this beetle is the hooding of the head by the 'prothorax' or chest. This is shown in the illustrations of the beetle (Figs. 11 and 12 and frontispiece). To a greater or lesser extent, this hooding of the head is shown by all members of the family ANOBIIDAE as well as by members of the related families PTINIDAE (the Spider beetles) and BOSTRICHIDAE (the Powder post beetles).

Fig. 13. Upper and lower surfaces of a floor-board; the lower showing the typical circular 'flight holes' of *Anobium punctatum*. Before lifting, the board was thought to be free from infestation!

Life cycle

The egg

From twenty to sixty eggs are laid by each fertilized female *Anobium*. The maximum possible is about eighty but the average is said to be twenty-eight. They are laid usually in small groups, in cracks of unpolished wood (Fig. 13), but sometimes they may be found just inside the old flight holes. In shape they are like oval pearls, but they are often pushed a little out of shape when being forced into crevices during egg deposition. They are easily seen by the naked eye and the illustration shows a group of eggs found on the plywood back of a picture frame. They are seen to be wedged underneath the rough splintery grain.

The larva

The young larvae emerge in about five weeks' time, not from the top of the egg but from the base, and immediately commence to tunnel their way into the wood (compare with the young larvae of Death watch). The larval stage is spent entirely in the wood and extends for not less than two years, the average length of the larval stage probably being something between three and four years; periods of five or even more are perhaps not uncommon. Out-of-doors an annual life cycle is said to take place.

The larva (see frontispiece) is greyish-white in colour and covered with fine hairs. The head is yellowish-brown, but the jaws are dark chestnut-brown. It usually holds its body in a hook-shape when in its tunnel, or tightly curved into a ball when it is brought out of the tunnel into the light.

The larvae remain entirely within the wood, tunnelling mostly up and down the grain, but often crossing into a different growth ring. Tunnelling takes place in cold weather as well as in summer, although at a decreased rate, and after tunnelling the larva withdraws a short distance from the 'working face' for an hour or two before recommencing operations. The tunnel which is left behind is loosely filled with powdery bore-dust which is known as 'frass'. The powder when rubbed into the palm of the hand is gritty, quite distinguishable from the powder of *Lyctus*, which is soft and silky (Fig. 51). If the frass of *Anobium* is viewed under a lens it is seen to consist of cigar-shaped pellets made up of minute fragments of the chewed wood. In severe infestations where the galleries intercommunicate large amounts of the frass may be forced out of the old flight holes. 'Powdering', however, does not occur to the same extent as in *Lyctus*.

The pupa

In the early spring of the year in which the larva matures, it bores towards the outside of the wood, but just short of the actual surface it constructs a pupal chamber slightly larger than the diameter of the gallery and changes into a pupa or chrysalis (Fig. 7). The pupa is creamy-white in colour and in many ways resembles the adult beetle. All the legs, wing-cases, antennae, etc., can be seen plainly, but are held quiescent by the thin transparent pupal skin. It can move only the last few segments of the abdomen, but in about six to eight weeks the pupal skin is burst off and the adult ecloses.

Although the adult stage has at last been reached a period is spent resting whilst the parts harden and turn to the characteristic brown colour. Then the beetle bites its way out into the open air and the hole it makes, which averages about 1·5 mm in diameter, is termed the 'flight hole' (Fig. 13). The adult beetle stage extends over two or three weeks only. During this time mating takes place (sometimes within a few hours of emergence) and then the female lays her eggs in the sort of places described and illustrated in figure 14. It is of interest, however, to mention that the females will stay in old flight holes, both before mating and afterwards, and the act of mating often takes place with the female beetle just inside the old flight hole with the tip of her abdomen just protruding and the male on the outside of the piece of wood. If insecticidal treatment of the piece of wood is carried out when a number of females are in hiding, they back out of the holes greatly disturbed, but the layman often gathers the impression that the fluid has caused them to bite their way out of the wood, which, of course, they have not done.

Fig. 14. Eggs of Common furniture beetle, seen here wedged in a crevice of birch plywood. Much enlarged.

Many wood-boring insects (and *Anobium punctatum* is one of them) have the power to bore through materials other than wood when endeavouring to escape from it on the assumption of the adult stage. Escape from painted and french-polished wood is, of course, very common and well known. In this case, if all the free surfaces of the wood are well painted then the process of painting must have taken place subsequent to the boring of young larvae into the wood. Linoleum and leather are two materials often perforated by the emerging adult Furniture beetle – the latter without difficulty, but linoleum does appear to kill a proportion of them as will be seen by an examination of the underside of perforated linoleum, when some dead beetles will usually be seen imprisoned in it. A variety of made-up goods lying in warehouses as well as books in libraries have been tunnelled into by emerging adults making their way out of wooden shelving.

Woods liable to attack
At this point no apology is made for giving some brief notes about the botany of wood, because without some elementary knowledge of timbers the significant differences between the biologies of the different wood-boring beetles would not be understood.

In the first place we have to distinguish between the timbers of the cone-bearing trees (Gymnosperms) and the broad-leaved trees (Angiosperms). The timbers of the former trees are known as softwoods, and the latter hardwoods. (Examples of softwoods are fir, pine, spruce, larch, and of hardwoods oak, elm, ash, birch, mahogany, walnut, obeche). It is important to understand this definition because today many tropical hardwoods which are now coming into large-scale use are actually softer than many softwoods and on the other hand there are several 'hard' softwoods.

In both types of timber, however, the wood is differentiated into the outer layer of sapwood and the inner core of heartwood. This is shown in the accompanying illustration (Fig. 15). In many timbers there is a great difference between heart and sap (such as laburnum where the heart is chocolate-brown and the sap canary yellow) but yet in others heart and sap may be identified only with difficulty. There is, however, this great biological significance; only the sapwood contains living tissue and only in this outer layer of wood is water transported throughout the trunk and branches. The heartwood consists of dead tissue only and into the heartwood is often excreted matter no longer required by the tree for the living processes. It is often this deposition of waste matter which alters the colour or the texture of the heartwood (and, of course, its durability and many other properties).

The subject of the types of wood which *Anobium punctatum* will infest is not a simple one, indeed there are several controversial points, but the following seems fairly certain, at least in the British Isles.

Softwoods. Historically, it seemed that only old and well-seasoned wood was attacked; then, in the 1950s, the author found that the sapwood of the softwood was attacked, but usually not before the timber had been

Fig. 15. Transverse sections of the trunk of a broadleaved tree (left) and a coniferous tree (right). The sapwood in each case contains living tissue but the heartwood consists only of dead tissue.

converted for some years. The heartwood of softwoods was attacked some time afterwards. Today, however, softwood is reported as being attacked by *Anobium punctatum* long before this time, although it appears that in such cases the attack proceeds at a fairly slow rate for a few years.

It does appear to the writer that the possibility of a genetic change taking place in the insect cannot be ruled out. Just as resistance to the action of chlorinated hydrocarbon insecticides has been bred into certain insect species, so the use of certain softwood species, perhaps from North America, where *Anobium punctatum* is not indigenous, is producing a race of this insect which thrives on new softwood.

Hardwoods. Here again the wood appears to require a certain degree of maturation before attack takes place, but this period is in some cases greater than is the case with softwoods. This has been generally recognized for a long time in the layman's mind – the presence of a few flight holes would provide ample proof of the genuineness of a piece of antique furniture. Beech appears to require a period of about forty years, and a large number of pianos of German origin, containing substantial quantities of beech, became infested with Furniture beetle after 40 to 60 years.

Some timbers are, however, anomalous in this respect. Birch is an example. Plywood made from birch is often found to be infested within four or five years of manufacture. Similar to birch in this respect are the timbers of alder and willow.

Almost all timbers are capable of being infested by *Anobium punctatum* but it has been observed by the writer that all timbers carrying a live infestation are without apparent fungal attack.

Let us now examine the increase of Furniture beetles: there seems to be no doubt that the increase is related to the vast amount of softwood which came into this country between 1920 and 1940 for the housing programmes. All this wood is now susceptible to attack and the laws of probability have thus made it very much easier for a female Furniture beetle to find suitable wood for egg-laying: the vast social changes in housing have tipped the balance in favour of this pest.

The following home-grown hardwoods are attacked by Common furniture beetle but those timber species where no information is available are also listed:

Alder	*Alnus glutinosa*	Box	*Buxus*
Apple	*Malus pumila*		*sempervirens*
Ash	*Fraxinus excelsior*	Cherry	*Prunus avium*
Aspen	*Populus tremula*	Horse Chestnut	*Aesculus*
Beech	*Fagus sylvatica*		*hippocastanum*
Silver Birch	*Betula pendula*	Sweet Chestnut	*Castanea sativa*
White Birch	*Betula pubescens*	Common Elm	*Ulmus procera*

Dutch Elm	*Ulmus hollandica*	Black Italian	
Wych Elm	*Ulmus glabra*	Poplar	*Populus serotina*
Holly	*Ilex aquifolium*	Black Poplar	*Populus nigra*
Hornbeam	*Carpinus betulus*	Grey Poplar	*Populus canescens*
Lime	*Tilia vulgaris*	White Poplar	*Populus alba*
Oak,		Robinia	*Robinia*
pedunculate	*Quercus robur*		*pseudoacacia*
Oak, sessile	*Quercus petraea*	Sycamore	*Acer*
Evergreen			*pseudoplatanus*
Oak	*Quercus ilex*	Walnut	*Juglans regia*
Turkey Oak	*Quercus cerris*	Cricket Bat	*Salix alba* var
Pear	*Pyrus communis*	Willow	*caerulea*
London		Crack Willow	*Salix fragilis*
Plane	*Platanus acerifolia*	White Willow	*Salix viridis*

Some authorities have stated that resin-bonded plywoods are not susceptible to attack. This is a generalization which is not true in particular cases. The writer has seen a bedroom suite in which all the resin-bonded plywood with which it was backed had been infested with *Anobium punctatum*. The explanation is this: before the sheets of plywood were separated from each other, eggs of *Anobium punctatum* were laid around the edge. The resulting larvae thus had two plies in which to develop which were not bonded but just closely pressed to each other. Where the larvae came into contact with the resin bond they turned back away from it. When the sheets of ply were separated the larvae would have fallen out.

The following home-grown softwoods are attacked by Common furniture beetle:

Atlas Cedar	*Cedrus atlantica*	Giant Fir	*Abies grandis*
Cedar of Lebanon	*Cedrus libani*	Larch	*Larix decidua*
Western Red Cedar	*Thuja plicata*	Japanese Larch	*Larix kaempferi*
Lawson's Cypress	*Chamaecyparis*	Corsican Pine	*Pinus nigra* var
	lawsoniana		*calabrica*
Deodar	*Cedrus deodara*	Scots Pine	*Pinus sylvestris*
Douglas Fir	*Pseudotsuga*	European Spruce	*Picea abies*
	taxifolia	Sitka Spruce	*Picea sitchensis*
Silver Fir	*Abies alba*	Yew	*Taxus baccata*
Noble Fir	*Abies nobilis*		

Some information is now available concerning a number of imported hardwood timber species. The following are susceptible to attack by the Common furniture beetle:

American Ash	*Fraxinus* sp.	Magnolia	*Magnolia acuminata*
Balsa	*Ochroma* sp.	Obeche	*Triplochiton*
			scleroxylon

Some species have been found to be moderately resistant to attack:

African Mahogany	*Khaya* sp.	Keruing	*Dipterocarpus* sp.
Agba	*Gossweileroden- dron balsami- ferum*	Japanese Oak Ramin Red Meranti	*Quercus* sp. *Gonystylus* sp. *Shorea pauciflora*
Central American Mahogany	*Swietenia* sp.		

The Common furniture beetle does not appear to be capable of attacking the following species which are immune or very resistant:

Abura	*Mitragyna ciliata*	Idigbo	*Terminalia ivorensis*
African Walnut	*Lovoa klaineana*	Iroko	*Chlorophora excelsa*
Afrormosia	*Afrormosia elata*	Sapele	*Entandrophragma cylindricum*

It is very interesting to note and certainly gratifying to the owners of Victorian furniture that Central American Mahogany, used extensively for furniture manufacture in the past, has been found to be moderately resistant to attack.

Wickerwork baskets (willow-twigs – entirely sapwood), broken and thrown into the loft, have very often been a reservoir of infestation for many years. They produce an annual crop of adult Furniture beetles which have laid their eggs on furniture and structural timbers throughout the house. The transport of infested furniture, suitcases, baskets and other items is probably a very important factor in the spread of infestations from one property to another.

Apart from the different timbers which are attacked, the question arises as to whether Furniture beetle prefers any special conditions such as temperature or disturbance. Many years of laboratory research would be required to give the answer to these questions, but the prevalence of infestations of Furniture beetles in the 'cupboard under the stairs' appears to the writer not to be a preference for undisturbed situations as such, but to a confined distribution. That is, when once the distribution has commenced (probably due to the birch plywood panelling in the first place), the adult beetles are not able to fly away from the closed cupboard-space and they thus lay their eggs back onto the already-infested woodwork. Thus a considerable infestation is built up relatively quickly. It is also conceivable that the preferred range of moisture content of the infested timber is a little above the normal – say 14 to 18% – a moisture content which would be quite likely in unventilated woodwork.

The parasites and predators of woodworm
Common furniture beetle has a number of parasites and quite a few predators preying upon it, some of which exhibit quite extraordinary

biologies. It might have been thought that wood-boring insects lived under somewhat sheltered conditions protected by the wood, but such is not the case. Populations of Common furniture beetle often decrease, indeed die out altogether, after a period of intense activity, especially when old oak sapwood is being attacked. Whether this is due to parasites, whether the heartwood is not susceptible to attack, or whether it is due to some other cause, is difficult to ascertain, but it is likely that parasites sometimes play a large part in reducing the activity of an *Anobium* infestation. On the other hand it is extremely unlikely that parasites are ever responsible for a complete extermination of wood-worm.

A most interesting feature of the great increase in Furniture beetle has been the corresponding increase of the hymenopterous parasite *Theocolax formiciformis* Westwood. (The HYMENOPTERA is the insect group containing ants, bees and wasps.) This small pteromalid wasp looks like a small black ant as described in its specific name (Fig. 16). It is often seen running in and out of the old flight holes of *Anobium* or its own flight holes, which are considerably smaller than the latter, being only about 0·5 mm in diameter. In almost all the adults the wings are rudimentary and even those that possess wings cannot fly. The female lays her eggs (maximum fourteen, average four to eight) on the back of the Furniture beetle larva or near to it, finding the larva by running through the *Anobium* tunnels or by cutting through the thin layers of wood with the ovipositor. The larval parasites devour their hosts fairly rapidly, completing this in about one to two weeks. The *Theocolax* larvae are described as ectoparasitic (i.e. they live on the surface of the *Anobium* larvae).

Fig. 16. The small (3·75 mm) blackish ant-like pteromalid wasp, *Theocolax formiciformis*, is a beneficial insect, its larvae being parasitic on the larvae of Common furniture beetle.

The pupal stage lasts about three weeks and the emerging adults make their way to the outside *via* an old *Anobium* flight hole or they may bite their own way out as previously mentioned. Three specimens once gnawed their way through half-inch corks and escaped in the writer's laboratory. The total length of the life cycle varies from about thirty-nine to seventy-five days according to climatic conditions so that emergence may take place throughout the year.

It has been found in the laboratory that, when gamma rays had been used to kill *Anobium* larvae, *Theocolax formiciformis* adults emerged apparently unharmed and similarly this species emerged safely from wood which had been subjected to V.H.F. heat treatment sufficient to kill *Anobium* larvae.

Another member of the group HYMENOPTERA parasitizing *Anobium* larvae is the little Braconid *Spathius exarator* (L.), and from 65 to 94% *Anobium* larvae have been reported as being infested with the insect. The adult *Spathius* is often observed walking rapidly over a window pane, obviously strongly attracted to light.

The beetle *Korynetes caeruleus* (de Geer) (Fig. 17), which is quite a common predator of Death watch beetle, also attacks Furniture beetle. The larva of this beautiful steely-blue beetle moves slowly around the intercommunicating galleries devouring the wood-eating larvae when it finds them. It has a superficial similarity to the larva of the Brown house moth (*Hofmannophila pseudospretella*), but it may be distinguished from the latter by the possession of a pair of blade-like processes at the hinder end of the body (Fig. 18).

A number of other species of beetle larvae are also recorded as being predacious on woodworm larvae, amongst which only the rather large *Opilo mollis* deserves mention. It is closely related to the aforementioned

Fig. 17. The adult beetle of *Korynetes caeruleus*. Length: 4 mm. This beneficial insect is steely-blue in colour.

Fig. 18. Predacious larva of the beetle *Korynetes caeruleus* (de Geer). This larva hunts out and feeds on the immature stages of Furniture beetle and Death watch. It is therefore a beneficial insect. The larva grows to 14 mm and is creamy-white with a chestnut-coloured head which can be retracted a long way into the first thoracic segment.

beetle *Korynetes caeruleus*, both being included in the beetle family Cleridae.

When infestations of *Anobium punctatum* and the Death watch beetle have been exterminated or controlled by fumigation or by the use of oil-soluble insecticides it often happens that *Korynetes caeruleus* beetles afterwards emerge unharmed.

Finally, we leave the insect group for perhaps the commonest parasite of Furniture beetle – the mite (a member of the spider group) *Pyemotes ventricosus* Newp. These are sometimes present in great numbers. They suck the juices of the *Anobium* larvae and the gorged females (Fig. 19) can be seen looking like shining beads on the larva's body, about one millimetre in diameter (Fig. 20). The males never leave the outside of the body of their mother (Fig. 19) and mate with their sisters as they are produced.

All these parasites and predators of woodworm are, of course, beneficial and should be preserved as far as possible. Certainly the expert in woodworm disinfestation should be able to recognize them. However, it should always be borne in mind that the total extermination of the host (in this case the woodworm larvae) is seldom, if ever, achieved by their help alone, and thus the occurrence of one or more of these parasites should never be a reason for deciding against insecticidal or other control measures.

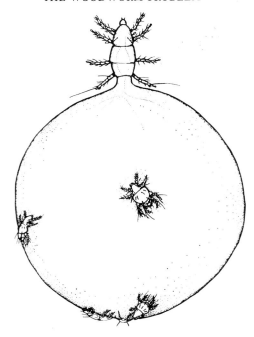

Fig. 19. A gorged female mite of the species *Pyemotes ventricosus* Newp. Length: 0·75 mm. This mite is of great interest from a biological point of view in that the male mites, which are very small, never leave the surface of the body of their mother. Here they mate with their sisters as they are produced. The young female mites as they are fertilized drop off their mother and seek out a host to parasitize. This may be a Furniture beetle larva or the larva of some other insect. *Pyemotes* mites probably represent a very large factor in the control of many insect pests. Greatly enlarged.

Fig. 20. Larva of Common furniture beetle infested with three *Pyemotes* mites.

3   Wood-boring weevils

Four weevils, beetles in the family CURCULIONIDAE, are wood-borers. All have a typical 'weevil' appearance with a long snout or 'rostrum' on which the antennae are situated, and a round cylindrical body and short legs. In this they have a strong superficial resemblance to the grain weevil *Sitophilus granarius*, but they can be distinguished from the latter insect by having the wing-cases completely covering the abdomen, whereas in the grain weevil the last segments of the abdomen are uncovered.

The four species concerned in boring into building timbers are *Euophryum confine* (Broun), *Euophryum rufum* (Broun), *Pentarthrum huttoni* Wollaston and *Caulotrupodes aeneopiceus* (Boheman) (previously, *Caulotrupis*). The last mentioned is rare in buildings and will be omitted from this review.

Euophryum confine
Sometimes called the New Zealand weevil. It was as recently as 1937 that this small beetle was first recorded as attacking wood in this country. In 1940 it was observed in a dead willow tree in Lea Valley Marshes, and in 1947 in Epping Forest. Later in 1947 it was reported by Dr. Hinton and Mr. McKenny-Hughes, of the British Museum, as attacking woodwork in London houses, and it was then pointed out that this weevil showed every sign of becoming a serious pest in houses.

These two entomologists were certainly right, as only a few years later it became widespread, not only in London and the adjoining counties, but throughout England. There seems to be no reason why this species should not extend over the British Isles; indeed, it may have done so already, as it has been reported in Dublin. It is, however, always found associated with wood in which there is evidence of decay, and the fungus usually responsible is *Coniophora puteana*, the cellar fungus. *Coniophora puteana* can be killed by drying the wood, but the New Zealand weevil will often continue to infest wood even after drying if *Coniophora* has once been in it.

The adult beetle of *Euophryum confine* is small, from 2·5 to 4·5 mm in length; on the average being a little smaller than the related *Pentar-*

Fig. 21. Adult *Euophryum confine*, a common wood-boring weevil.
Length: 3·8 mm.

thrum huttoni. It is blackish-brown to reddish-brown in colour and the
typical long snout or rostrum can be seen quite well with the naked eye
(Fig. 21). It walks with rather a 'loping' gait.

Unlike most of the other commonly-occurring wood-boring beetles
the adult stage of this species burrows on its own account, and is often
found quite deep in the wood. The adults can be found almost through-

Fig. 22. Wood damaged by a weevil. Note irregular flight holes and surface
channelling caused by the adults.

out the year, the reason being that there seem to be two overlapping life-cycles in the year. Also, the adults may live for as long as 16 months.

The flight holes average 1·1 mm in diameter – that is, definitely smaller than the flight holes of *Anobium punctatum* and only the smallest of the *Lyctus* flight holes approach it in size. *Euophryum* flight holes, however, are always ragged in outline (Fig. 22), very distinct from *Anobium* and *Lyctus*, and often a sprinkling of frass (Fig. 51) occurs around the margin of the flight hole. Curiously enough this is a case of an insect of no importance in its country of origin becoming a pest of economic importance when introduced into another.

Euophryum rufum

This weevil is very similar in appearance to the last and indeed, there is some doubt as to whether it is a distinct species. So far as is known, its habits too are very much the same as *E. confine*.

Pentarthrum huttoni

In general appearance this species is very like the foregoing. It is, however, more blackish and not nearly so reddish in colour as *Euophryum confine*. The larva and adult bore into timber which is damp and infected with one of the wood-rotting fungi. In addition, however, it is sometimes found in very dry timber, but in these cases it is always evident that some time in the past the timber has been wet and partially rotted by a fungus. Wherever such wood is found, *Pentarthrum huttoni* may be present – it is commonly found in woodwork near leaking water-pipes in the vicinity of sinks, lavatories, bathrooms and similar situations, but it is amazing how often attacks by this beetle are found in breweries, wine vaults and beer cellars. They certainly seem to thrive in such places!

Fig. 23. The larvae of *Pentarthrum huttoni* and the other wood-boring weevils are entirely legless. Length: 3·3 mm. Taken from wood rotted by the fungus *Coniophora puteana*.

Fig. 24. A pupa of *Pentarthrum huttoni.* Length: 3·5 mm.

Very little is known about the duration of the different life-stages (Figs. 23 and 24) but hardwoods and softwoods can be attacked, and damp plywood used as panelling against a wall is especially favoured. One curious characteristic is often observed concerning the appearance of the attacked wood. The softer layers of the wood are eaten entirely away first, leaving the harder rings untouched. This leaves the wood with a curious appearance, often like that of termite-attacked wood in the tropics.

4 Death watch beetle

The Death watch beetle is an indigenous British insect which, on account of its habits, has been well known for several hundred years. It was given the name *Xestobium rufovillosum* by De Geer in 1774, which is the name now generally used, although the name *tessellatum*, given to it by de Villers in 1789, is found in much older literature. Its outdoor habitats consist of dead wood in trees or dead branches of several hardwood species where fungal decay occurs, a common situation being the dead wood in the trunk or crown of pollard willows. Oak, ash and sweet chestnut are also commonly infested by this species, and hornbeam, poplar and whitethorn more rarely.

The adult beetle is usually 5 to 7 mm in length, the females being usually slightly the larger. It is dark greyish-brown in colour, with a pattern of yellowish scale-like hairs on the pronotum and wing-cases.

Fig. 25. Death watch beetle, *Xestobium rufovillosum*. Length: 6 mm.

The wing-cases, however, often become rubbed when the colour may be more reddish and shining. The longitudinal rows of small pits on the wing-cases, present in *Anobium punctatum*, are absent in Death watch (Fig. 25).

The pronotum, which is much more widely flanged, hoods over the head just as in Common furniture beetle (Death watch beetle is in the same family – the ANOBIIDAE – as Common furniture beetle *Anobium punctatum*).

The tapping of the adult beetles is a well-known phenomenon. Both sexes tap, and it is the sound of the tapping in the quiet hours of the night in a sick room which has given it its common name.

In captivity the adults can be stimulated to tap by four or five sharp staccato raps with a pencil. The tapping is believed to be of value in finding mates. When the Death watch is about to tap, it appears to stiffen and to hold the front of the body away from the wooden surface with the tip of the abdomen touching the surface. It then strikes its head against the wood about twelve times within a second and a quarter, the whole body of the beetle moving.

Life cycle

The egg

The eggs are laid in small clusters, usually of three or four, on the surface of rough wood, cracks or just inside the lip of flight holes. They are sticky when first laid and adhere to each other. The eggs are whitish, oval and much like those of the Common furniture beetle, but are, however, twice as large. Forty to sixty is the average number of eggs laid, but up to 201 have been recorded. The length of this stage varies from about two to five weeks.

The larva

Again, the larval stage has the same general appearance as in Common furniture beetle. A point of difference, however, is found in the eyes. Whereas in Common furniture beetle these consist of one black spot on each side of the head, in Death watch two black spots occur on each side. Another difference, this time in behaviour, is also apparent. When the young Death watch larvae hatch out from the eggs (Fig. 26) they walk about quite rapidly over the surface of the wood before finding an old flight hole or crevice through which to enter the wood. The larva is creamish-white, strongly hook-shaped when taken out of its tunnel, covered with erect golden hairs and grows to a length of about 10 mm.

The length of the larval stage is very variable. Under the best possible conditions for growth which can be obtained in the laboratory – continued summer warmth, high relative humidity, and with well decayed willow as foodstuff – the life cycle can be completed in a year. Under less favourable conditions it may not be completed until ten years. It is,

Fig. 26. Larva of the Death watch beetle leaving its egg. It walks actively over the surface before starting to tunnel into the wood. This contrasts with the larva of the Common furniture beetle, which bores through the underside of the egg and tunnels straight into the wood.

however, generally much longer out-of-doors: in fungal infected oak it has been found to vary from three to seven years and it is considered that an average of about four and a half years as the larval period indoors would not be unreasonably far from the truth.

The frass of Death watch is quite characteristic; it can be seen clearly with the naked eye to consist of bun-like pellets (Fig. 51).

The pupa
This is much like that of *Anobium punctatum*. It is milky-white in colour at first but then darkens progressively.

During the summer in which it reaches full growth, the larva burrows its way towards the outer surface of the wood which it is infesting and constructs a pupal chamber by enlarging the larval gallery and plugging the open end with frass. It changes into the pupal stage in late summer, remaining three or four weeks in this stage, and then changes into the adult beetle but remains inside the pupal chamber without biting its way out until Spring. The usual period of emergence is during the latter part of April and the beginning of May. The earliest date recorded is 24 March and the latest 1 August.

Fig. 27. A boss from Burwell Church, Cambridgeshire, as seen from the ground and showing (above) only slight evidence of Death watch attack. The reverse (below) shows the almost complete destruction of the back of the boss.

The flight holes are circular and approximately 2·5 mm in diameter. The pattern of distribution of the flight holes often follows the course of the fungus-infected wood to a remarkable extent.

Damage to buildings

Indoors, Death watch infests hardwood structural timber which has at one time or another maintained a fungal infection. Oak is the most common timber attacked, but chestnut, elm, alder and walnut are sometimes infested. On occasions, however, softwood is attacked if it is adjacent to infested hardwood but attacks only originate in hardwood.

The Death watch infested timber can appear quite sound (Fig. 27) yet the rich chocolate-brown colour of the tunnelled timber make it very probable that the living tree from which the timber was converted was attacked by the beef-steak fungus, *Fistulina hepatica*. This is the fungus which produces 'brown oak'. In other instances the cellar fungus *Coniophora puteana*, causing a wet rot, and *Donkioporia expansa (Phellinus cryptarum)* and *Coriolus (Polystictus) versicolor*, causing white rots, have brought about the incipient breakdown of the wood required by the Death watch beetle. In other words, the larva seems to require the wood to be partially predigested for it.

This phenomenon is a common occurrence amongst the wood-boring beetles, but on the other hand the larvae of many of the common species such as Common furniture beetle (*Anobium punctatum*), *Lyctus* species and House longhorn (*Hylotrupes bajulus*) feed on sound wood in buildings where fungal infection is apparently absent.

Death watch beetles rarely fly (see p. 22) and this raises the question as to how they come to infest timber indoors. It seems virtually certain that Death watch infestations in buildings were initiated by the use of timber of large dimensions already infested when the tree was growing and few new infestations take place today. This means that with the current work of control the total number of infestations indoors is decreasing.

5 Powder post beetles

Six species of the *Lyctus* genus in the beetle family LYCTIDAE are found in Britain. The names of these species are as follows: *Lyctus brunneus* Stephens; *Lyctus sinensis* Lesne; *Lyctus linearis* (Goeze); *Lyctus planicollis* Leconte; *Lyctus cavicollis* Leconte; *Trogoxylon parallelopipedum* (Melsheimer). In addition to these six species *Lyctus africanus* is occasionally imported from various parts of the world.

Only the species *L. linearis*, however, ranks as a truly indigenous species. *Lyctus brunneus* (Fig. 28) was introduced into Britain many years ago from North America, so that today it is common and widespread. The other species, although now established in Britain, have

Fig. 28. *Lyctus brunneus*. Note head distinctly visible from the top (cf. *Anobium*). *Lyctus* is also longer and thinner than *Anobium punctatum*. Length: 5·5 mm.

46

Fig. 29. Diagrammatic view of female *Lyctus* beetle laying eggs.

only been introduced in recent years. They are very much less common than *L. brunneus*.

The egg of *Lyctus* is long and maggot-like, approximately cylindrical, but tapering off to a tail-like thread. The female beetle places her eggs with care and precision in the early wood vessels or pores (Fig. 29). These may be exposed in the sectioned wood as cut or indeed opened by the beetle herself in the 'tasting' marks. They are placed nowhere else, and this is a strong, if mechanical, reason why *Lyctus* damage is restricted to the so-called 'wide-pored' hardwoods. Up to fifty eggs are laid.

When the egg hatches, the young larva finds itself already in a pore, but is smaller in diameter than the pore. It then feeds on a yolk-like substance left behind in the egg until its girth is sufficient to enable it to move down the pore by gripping the sides with its body. It makes its way along the pore and finally pierces it. The larva then bores its way backwards and forwards in the sapwood, gradually increasing in size to as

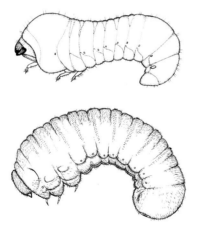

Fig. 30. Larva of (above) *Lyctus* sp. and (below) *Anobium punctatum*. Note that the terminal spiracle (or breathing pore) is large in *Lyctus* and small in *Anobium*, whilst head of *Lyctus* is embedded in thorax to a greater degree than in *Anobium*.

much as 5 mm in length. It is creamy-white, hook-shaped and segmented rather prominently. The head is light creamy-yellow with the jaws dark chestnut. The legs are very small, but a distinctive feature of the *Lyctus* larva is the pair of spiracles or breathing pores on the last abdominal segment, which are large and rusty red in colour (Fig. 30).

The frass, or bore-dust, of Powder post beetles is very fine and looks like flour (Fig. 51). The larvae gradually reduce the timber to dust, leaving a thin veneer of sound wood on the outside. The infested article then crumbles at a touch, hence the name Powder post. Flight holes are about 1·4 mm in diameter.

Starch content of the wood

The *Lyctus* larva feeds on the cell contents, containing mostly starch, sugars and related substances together with a little protein. The cell walls cannot be digested. A moisture content in the wood of 8 to 30% is necessary for *Lyctus* attack.

The essential part played by the starch content in the initiation of *Lyctus* attack is well known; a 3% starch content of sapwood of wide-pored hardwoods is required by female *Lyctus* beetles before they will lay their eggs.

It has been suggested that a substance other than starch, but which is water-soluble, is the substance which excites the female to lay her eggs, this being the explanation of the fact that fresh-cut timber which has been thoroughly washed in water is immune from *Lyctus* attack.

Lyctus is usually considered as a pest of new wood; that is, wood in its first few seasons after felling – whilst the starch content of the sapwood is at the required concentration. There does, however, seem quite an amount of evidence to suggest that timber very much older than this is attacked by *Lyctus*. Oak flooring strips in various forms are a case in point. The only explanation of this phenomenon appears to be that kiln-drying sometimes fixes the starch content.

Timbers attacked

The following home-grown hardwoods are attacked by *Lyctus* beetles:

Ash	*Fraxinus excelsior*	Evergreen Oak	*Quercus ilex*
Sweet Chestnut	*Castanea sativa*	Turkey Oak	*Quercus cerris*
Common Elm	*Ulmus procera*	Robinia	*Robinia pseudoacacia*
Dutch Elm	*Ulmus hollandica*	Sycamore	*Acer pseudoplatanus*
Wych Elm	*Ulmus glabra*	Walnut	*Juglans regia*
Hornbeam	*Carpinus betulus*	Cricket Bat Willow	*Salix alba* var *caerulea*
Pedunculate Oak	*Quercus robur*	Crack Willow	*Salix fragilis*
Sessile Oak	*Quercus petraea*	White Willow	*Salix viridis*

Note: NO SOFTWOODS ARE ATTACKED BY THESE INSECTS, although a single case has been recorded of an infestation in a piece of starchy softwood – a high starch content in softwood is unusual.

Here is a list of imported timbers known to be susceptible to attack by *Lyctus* beetles:

Abura	*Mitragyna ciliata*
Afara	*Terminalia superba*
Agba	*Gossweilerodendron balsamiferum*
Black Bean	*Costanospermum australe*
Australian Blackwood	*Acacia melanoxylon*
White Bombay	*Terminalia procera*
Borneo Camphorwood (Kapur)	*Dryobalanops* spp.
African Celtis	*Celtis soyauxii*
White Chuglam	*Terminalia bialata*
Crabwood	*Carapa guianensis*
Dhup	*Canarium euphyllum*
Eng	*Dipterocarpus tuberculatus* and *D. obtusifolius*
Black Guarea	*Guarea thompsonii* probably susceptible
Spotted Gum	*Eucalyptus maculata*
Gurjun	*Dipterocarpus* spp.
Idigbo	*Terminalia ivorensis*
Iroko	*Chlorophora excelsa*
Ironbark	*Eucalyptus* spp.
Jelutong	*Dyera costulata*
Kokko	*Albizzia lebbek*
Indian Laurel	*Terminalia* spp.
African Mahogany	*Khaya ivorensis et al.*
Muninga	*Pterocarpus angolensis*
Tasmanian Oak	*Eucalyptus obliqua, E. regnans* and *E. gigantea*
Obeche	*Triplochiton scleroxylon*
Amboyna	*Pterocarpus indicus*
Indian Rosewood	*Dalbergia latifolia*
Sapele	*Entandrophragma cylindricum*
Seraya, Meranti	*Shorea* spp. and *Parashorea* spp. *et al.*
Teak	*Tectona grandis*
African Walnut	*Lovoa klaineana*

Bostrichid beetles

Rather closely related to the *Lyctus* beetles is the family BOSTRICHIDAE, which contains a large number of mostly exotic Powder post beetles. In general biology they resemble the *Lyctus* group in that they are only found in the sapwood of hardwoods rich in starch. Most species are larger than *Lyctus*, some being up to 25 mm in length. In shape they are cylindrical and have something of the hooded prothorax of the Anobiid beetle, but contrasting with the latter in that the prothorax is curiously beset with spines. Bostrichid beetles are essentially pests of the freshly-felled tree, and under the conditions of the timber-yard they quickly die out. They do not re-infest the dried timber, but some species are able to continue their development in dry and fabricated woodwork and will sometimes emerge from furniture and ornaments.

Nevertheless, Bostrichid beetles are a source of annoyance and loss to the timber importer, as often logs are received from abroad with all sapwood completely unusable on account of the large tunnels.

6 Longhorn beetles

The CERAMBYCIDAE is a family of beetles of widespread distribution throughout the world. They are popularly known as the 'Longhorns' and about 60 species regularly breed in the British Isles. In addition, many species from abroad are regularly brought into this country in the immature stages in imported timber. They are mostly medium to large-sized beetles and are characterized by having long antennae, in many cases extraordinarily long. The larvae of practically all of them are wood-borers. Some species are found in the sapwood of living trees, some in the bast (the layer immediately beneath the corky layer), yet others are found only in sound but dead sapwood, whilst others are found in wood already infested with various species of fungus.

Quite a number of species of CERAMBYCIDAE occasionally make their way out of furniture, structural timbers, pallets or packing cases, either because the larva develops naturally in dry wood or because, developing naturally in living sapwood, it is nevertheless able to complete its life cycle in dry wood. In this case the larval period is prolonged over many years. A large number of species of Longhorn beetles occur in sufficient numbers in various parts of the world to be of economic importance.

House longhorn beetle

One species, the House longhorn beetle, *Hylotrupes bajulus* (L.), shows a unique feature in its biology: alone of all the British Longhorns it is able not only to complete its development in dried timber, but to infest and re-infest old dried timber. It is probable that many House longhorn infestations are due in the first place to timber of European origin imported with the young larvae already burrowing in the wood. A number of records have been made of House longhorn larvae being found in packing cases and orange boxes. This may perhaps account for that part of its distribution which is scattered and discontinuous. With regard, however, to its local abundance in north-west Surrey, where, in some small towns and villages, a large proportion of the buildings has infested woodwork, it is certain that some combination of ecological features is favourable to its continued presence. It has recently been found that this circumscribed area is the warmest area in Britain. It is perhaps also significant that the areas in north-west Surrey where it occurs are sparse coniferous heathland with sandy soil, very similar to areas in Continental Europe which the insect inhabits.

House longhorn attacks only softwoods and, at first, the attack is confined to the sapwood, but nevertheless heartwood also is often found to be attacked at a much later stage.

In appearance the adult beetle is very variable in size, from 12 mm (males) to fully 25 mm in length (females). It is greyish-black to a very dark brownish-black in colour with two greyish transverse marks across the wing cases. Almost all the body is covered with greyish or yellowish-grey hairs, but the extent of this hairiness is variable. Some areas have the appearance of having the hairs rubbed off. A pair of such areas on the pronotum (the round part next to the head) have a remarkable resemblance to eyes (Fig. 31). The antennae are long but not unduly so.

Eggs of the beetle are yellowish to greyish-white in colour, elliptical with rather pointed ends, and are laid in cracks and crevices on the surface of the wood. The length of the egg stage is said to be about three weeks.

The larva is greyish-white and fleshy, with very pronounced segmentation (Fig. 32). The head is small with the greater part of it retracted into the body. The first three or so segments are large, the body then tapers, but the last two segments show an increase in size. The length of the full-grown larva varies from about 20 to 37 mm. When newly hatched, the larva bores straight into the wood. After a time the larval

Fig. 31. House longhorn beetle, *Hylotrupes bajulus*. Note eye-like spots on the pronotum (the division next to the head). Length: 12–25 mm.

Fig. 32. The larva of the House longhorn beetle. Note the deep inter-segmental grooves and absence of legs.

tunnel takes up an elliptical cross-section and large quantities of bore-dust are produced. The particles of bore-dust are not uniform. Some particles are very small, others are larger and have a rather granulated cylindrical appearance (Fig. 51), but the frass is so mixed with the air that it occupies a greater volume than the wood from which it is produced. This often is of importance in detecting an infestation because the packed frass causes the surface of the wood to give a blistered or rippled appearance.

When an infestation has been proceeding for some years, almost all the sapwood is converted to powder except for a thin veneer of surface skin. At this stage the surface skin often bursts along the grain (or where it was!).

Pupation and emergence

Pupation in nearly all the wood-boring Longhorn beetles follows approximately the same pattern. The larva eats its way close to the surface of the wood and then moves away again. It then blocks the gallery on the inner side, the particles of wood forming the plug being very coarse. The larva then changes to a pupa which is at first creamish-white but later becomes light brown. All the parts of the adult beetle can be made out in the pupa, but it is quiescent, not being able to move its legs. The abdomen, however, can be twisted about quite vigorously. The length of the pupal stage is thought to be about three weeks. When metamorphosis into the adult beetle has taken place the beetle bites its way out of the wood.

Fig. 33. Typical damage to the sapwood of softwood by the larvae of the
House longhorn beetle.

Flight hole
The flight hole is oval, the longer diameter varying from about 3 to
6 mm (Fig. 33). A relatively small number of larvae can cause a large
amount of damage to softwood joists and rafters. Six larvae can convert
to powder all the sapwood in a common rafter. Do not, therefore, expect
to find a large number of holes in a House longhorn infestation. A
phenomenon exhibited by the larvae is the extraordinary straightness of
the gallery when they find themselves in a new piece of timber. On one
occasion when an infested joist was cut down one gallery was found to
vary only 10 mm from the straight in a 3-metre run.

It is remarkable how often House longhorn infestations occur in the
roof space of a house, and this is probably due to the peculiar micro-
climate – very hot in summer and very cold in winter.

Often one is asked to identify an infestation in the laboratory from a
small splinter of wood showing perhaps a small part of a larval gallery
with no stages of the insect present. If the surface of the gallery is
examined critically with the naked eye the bite marks of the larvae can
be observed. The appearance is that of ripples in the sand left by the
retreating tide. Several incorrect identifications have occurred, however,
since numerous other Longhorns make the same ripple marks.

Survey
It is unfortunate for the householder that House longhorn beetle infestations seldom come to light before a great deal of damage has taken
place. The roof space is seldom visited – perhaps only once a year or so

Fig. 34. *Phymatodes testaceus* (L). Two forms of this beetle are found; one fulvous brown, the other with the wing-cases a steely blue-black. It is about 12–15 mm in length.

to stow away junk. This means that the extent of the damage should always be surveyed by an expert, as some of the structural timbers may have to be replaced or reinforced. Notes on control measures are given in Chapter 10.

The Oak longhorn   *Phymatodes testaceus*
Perhaps the most common Longhorn beetle met with in timber-yards is the species *Phymatodes testaceus* (Fig. 34). The fat, 25 mm long, legless grubs are commonly found in the bark and in the superficial sapwood. This again is a species which can complete its development in the dried and fabricated timber. When it is found in buildings it is usually infesting new oak flooring. The beetle is extremely variable in colour from yellowish-brown to deep blue-black.

The Wasp beetle   *Clytus arietis*
This beetle not only has the black and yellow-banded appearance of a wasp (Fig. 35), but by its agile, quick movements and its waving antennae it shows an extraordinary similarity to this insect. It is a common insect in the garden, where the larvae inhabit rustic ware and fencing, but is sometimes found emerging from furniture, which it will not re-infest. A number of closely related beetles, similar to *Clytus* in appearance, are known, from furniture manufactured from timber of foreign origin.

   The closely related species *Anaglyptus mysticus* is another British species which can delay its development over a number of years when

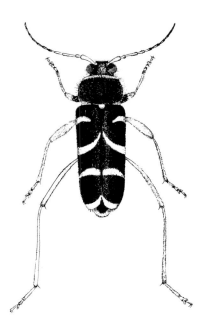

Fig. 35. The Wasp beetle, *Clytus arietis* (Linnaeus). Between 12 and 20 mm in length, this agile beetle often emerges from beech and oak furniture. The black and yellow bands give it a startling appearance.

the wood into which it is boring is made into furniture. Records of up to thirteen years are known. It is somewhat like *Clytus arietis* but it has large mahogany-coloured patches at the front of the wing-cases.

Arhopalus ferus

A British insect common in some southern areas is *Arhopalus ferus*. It is about the same size as *Hylotrupes bajulus* and is sometimes confused with the latter insect. The drawing (Fig. 36) shows, however, that this beetle does not have the shining spots on the thorax, nor does it have any white hairs on the elytra.

Although occasionally this species will attack apparently healthy trees, more usually the infestation occurs in decaying or dead stumps, dead standing trees (especially those scorched by fire) and cut logs in timber yards. Adults emerge from July to August and fly readily on warm days.

The species is of economic importance since it sometimes infests the softwood pallets used in many industries to carry stacks of goods. Damage can be so severe as to cause collapse of the pallets.

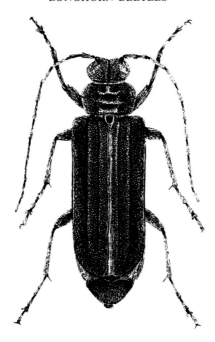

Fig. 36. *Arhopalus ferus.* The beetle is about 20 mm in length and is a dark brown colour.

The Basket beetles

Two rather small Longhorn beetles – *Leptideela brevipennis* and *Gracilia minuta* – are sometimes found infesting baskets and other articles made of unbarked willow and hazel twigs. The small flight holes – about 2 mm across – are distinctly oval, and the larvae are very similar to typical longhorn larvae, except that they are very small.

7 Some other wood-boring beetles

The Wharf borer *Nacerdes melanura*
This beetle is thought to be native to the Great Lakes region of North America and as it does much damage to piles and dock timber in this region it is known in the United States, and now in Britain, as the 'Wharf borer'. It is in the beetle family OEDEMERIDAE, a family with only seven British species. In Britain, the Wharf borer is widespread although more common in the South. The large greyish-white larvae live in very wet, rotten wood. The 'normal' habitat is at, or just above, the water line in river estuaries in piling, wharfs and timber supporting the river banks. In addition Wharf borers occur in wooden-hulled boats and also in marine situations in harbours: the amount of salt in the water apparently not affecting the larvae.

The range of suitable breeding sites has further increased in recent times to include timber inland which is soaked with rain water. During the 1940's Wharf borers became increasingly common in London due to the large amount of wet timber rotting in bomb sites. Since those times, however, this insect has spread to many other areas and now seems to thrive on buried builders' refuse. The wood, sometimes buried at a considerable depth, has been found to contain many living larvae years after it was buried. The adults can mate and lay their eggs within the cavities in the wood gnawed out by the larvae. Thus tracing the source of an infestation is rarely easy and often all that can be done is to kill the adult beetles as soon as they make their appearance.

The larva of Wharf borer (Fig. 37) cannot possibly be confused with any other of the wood-boring larvae we have described. It is often 25 mm in length and has a large yellow head. The six legs are also larger than hitherto seen in wood-boring larvae; indeed, in most cases larval legs are not well-developed. Paired patches of spinules occur on the top surface of the first five segments behind the head and on the underside of the sixth and seventh segments are pairs of stump-like false legs, also with spinules.

Well-defined tunnels are not constructed by the larva, nor does there appear to be much frass produced, but at intervals rather coarse wood

58

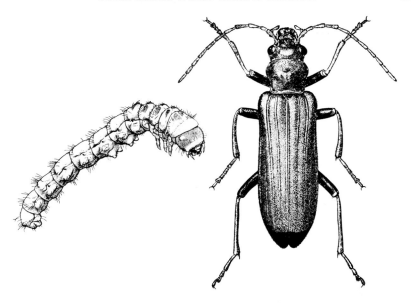

Fig. 37. The Wharf borer *Nacerdes melanura*. Light fulvous brown in colour with black-tipped wing-cases. Length: 7–14 mm. The large head and patches of spines are distinctive features of the larva. Reproduced by permission of the Trustees of the British Museum.

Fig. 38. Piece of rotten wood attacked by Wharf borers. Note the coarse wood scrapings in the tunnels.

scrapings are found which have not passed through the body of the larva (Fig. 38). Some infestations have been confused with House longhorn by the inexpert, as there is a similarity to old softwood well-tunnelled by the larvae of this latter species. There are, however, no grounds for confusion, as the inside walls of Wharf borer galleries never display the ripple appearance of the bite-marks always shown by House longhorn.

All wet timber attacked by Wharf borer larvae is infested with fungal hyphae – either Dry rot (*Serpula lacrymans*) or one of the other fungi causing rapid decay of wood where permanently wet conditions are concerned, such as *Coniophora puteana*.

The adult beetle (Fig. 37) is variable in size from about 7 to 14 mm in length. It is a soft flimsy insect, golden-brown to reddish-brown, with the tips of the wing-cases black (*melanura* means 'black-tailed') and with long antennae. There are three raised longitudinal lines on each wing-case – a feature common to all beetles in the family OEDEMERIDAE. (It is sometimes confused with the typical brown form of the Oak longhorn beetle, *Phymatodes testaceus* and with *Rhagonycha fulva*, a very common and widely-distributed beetle with many local names.) In some parts of London the insect has appeared in incredible numbers – tens of thousands running over the pavements of the Strand, and yet their origin remained a mystery. It has a strong flight.

Ernobius mollis

Another species in the family ANOBIIDAE is *Ernobius mollis* – a species without a common name. It is rather like a large Common furniture beetle, being up to 6 mm in length (Fig. 39). When freshly emerged it is covered with short golden hairs which give it a light shade of golden-brown which darkens when some of the light-coloured hairs are lost. It lacks the longitudinal rows of dark marks or 'punctures' on the wing-cases which are present on *Anobium punctatum*. The wing-cases and other parts are much less horny than in the Common furniture beetle and in other species in the family ANOBIIDAE, and its trivial name *mollis* refers to this characteristic, meaning 'soft'.

The eggs of *Ernobius mollis* are laid in summer in the bark of softwoods only, and the larvae, which resemble those of Death watch, thereafter tunnel in the bark and the outer ring of sapwood. Often this causes the bark to fall off, exposing the tunnels packed with coarse frass; where the larva has been feeding in the bark the frass is the same dark colour as the bark; where it has been feeding on the sapwood it is light in colour (Fig. 51). After the pupal stage has been passed the adult beetle bores out of the bark, leaving a round flight hole of approximately 2·5 mm in diameter. Hardwoods are never attacked. Softwoods are only attacked when the bark is still attached. This does occur perhaps more often in outbuildings and barns than in well-constructed dwellings, but even so, *Ernobius mollis* is known to occur in the latter. Softwood studdings, with bits of bark still attached, on which has been fixed ply-

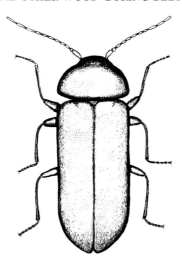

Fig. 39. *Ernobius mollis* is closely related to the Common furniture beetle but lacks the longitudinal rows of 'punctures' on the wing-cases. Length: 4 mm.

wood panelling, sometimes support infestations and the escaping beetles will bore through the panelling. It will be seen that little structural damage can be caused by this insect but sometimes anxiety is aroused when it is confused with Common furniture beetle. Any damage to timber would be superficial only. Control is easily effected by stripping off the bark where it occurs, when the infestation will rapidly die out.

Ptilinus pectinicornis
This is a small dark-brown beetle from 3·3 to 5·5 mm in length; it has a longer, more cylindrical body than Common furniture beetle, and has a curious, rather globular, prothorax, but the easiest distinguishing feature is the comb-like antennae of the male (Fig. 40), a feature not so exaggerated in the female, where the antennae are merely toothed like a saw.

In *Ptilinus* as in *Ernobius mollis* the longitudinal rows of 'punctures' on the wing-cases are absent. The larvae are found in relatively few timbers compared with Common furniture beetle; beech, maple, ash and sycamore being the most frequently attacked, but it is sometimes found in a mixed infestation with *A. punctatum*. Out-of-doors, however, the range of timbers that it will attack seems to be wider, but it is the hardwoods that are always attacked – softwoods are but rarely bored by *Ptilinus*. The flight holes are sometimes larger that those of *Anobium* but not so large as those of *Xestobium*. The 'frass' or bore-dust is fine and silky to the touch, reminiscent of *Lyctus*, again quite unlike that usually found in the ANOBIIDAE. Indeed, recent work on the biology of

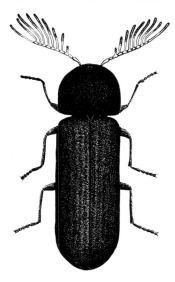

Fig. 40. Male adult of *Ptilinus pectinicornis*. The unmistakable feathery antennae are clearly shown. Length: 5 mm.

Ptilinus indicates that it may be more nearly related to the Powder post beetles than to *Anobium punctatum*.

The Golden buprestid   *Buprestis aurulenta* L.
Another wood-boring beetle which displays the phenomenon of delayed development is the species *Buprestis aurulenta* in the family BUPRESTIDAE – quite distinct from the Longhorn family (the CERAMBYCIDAE). This beetle is about 20 mm long and of a beautiful metallic green colour edged with copper. It is of North American origin, and when found in the United Kingdom it has usually emerged from Douglas fir or other softwood used in house construction such as for a door or window sill. Normally, the larva is found in the living tree, but nevertheless is able to complete its development in the seasoned and fabricated timber. In these circumstances development can be prolonged for several years – as long as 25 years has been recorded.

Pinhole and Shothole borers
Damage to timber by Pinhole and Shothole borers, or Ambrosia beetles as they are called, is one of the most serious problems confronting the widespread utilization of many tropical timbers. Quite a considerable proportion of these hardwoods (although the problem is by no means confined to the tropics nor to hardwoods) is damaged by these beetles. Unfortunately, a popular and widespread misconception exists in the

public mind against the use of such damaged timber, as confusion with 'woodworm' invariably exists. The old beliefs (although fallacious) die hard – 'There is no "cure" for woodworm and every article showing the tell-tale flight holes must be burnt'. Thus timber showing pinhole damage is looked upon with the gravest suspicion. It is also surprising that many timber merchants have failed to grasp the significance of the problems involved, and little if anything has been attempted to overcome the difficulties. On the other hand much valuable scientific knowledge has been gained by the work carried out in recent years by the West African Timber Borer Research Unit. The biologies of a number of species have been worked out so that we know much more about the insects and their mode of life and, in addition, much experimental work on control has been undertaken. A relatively large number of beetles, mainly classified in the families PLATYPODIDAE and SCOLYTIDAE (the chief genera being *Platypus, Diapus, Trypodendron, Anisandrus* and *Xyleborus*), have larvae which infest the trunks and branches of, in most cases, freshly felled trees, but sometimes living trees. Until recently little detailed knowledge existed concerning the biology of these insects, and there are differences in detail among the various species.

As an example of the life history of this group, that of the Oak pinhole borer, a British Ambrosia beetle (Fig. 41) is given here. The male beetle first flies onto the trunk of a freshly felled oak tree, probably attracted

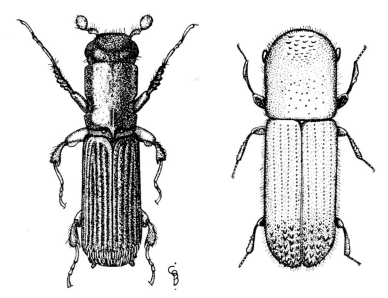

Fig. 41. Left: the Oak pinhole borer, *Platypus cylindrus*. Length: 5 mm. Right: *Xyleborus saxeseni*, another pinhole borer which attacks a variety of hardwoods and softwoods. Length: 2·2 mm.

Fig. 42. Damage to timber by a Shothole Borer, characterized by the dark staining of the wood around the holes.

by the smell of fermenting sap. He then constructs a tunnel, and it is to this that the female is attracted. After mating, the female carries on with the boring. She lays small batches of eggs at intervals which hatch into slug-like larvae. The latter feed on a fungal growth occurring on wood fragments and tunnel lining.

Staining fungi

A characteristic of the species of fungi which grow in close association with the tunnelling larvae is their property of staining the wood in the area where the fungus is alive. Purplish-brown is the usual colour of the stain, but it may vary from a relatively light hue to almost black. Sometimes the area of staining is confined to the walls of the tunnels, at other times the staining is more extensive – extending to fair-sized patches or flame-shaped areas (Fig. 42). This fungus has been termed 'Ambrosia' and is the reason why the pinhole beetles are labelled 'Ambrosia beetles'. Another characteristic shown by both pinhole and shothole is the absence of bore-dust. The tunnels appear quite free of any 'powder', so different from *Lyctus*, which shows abundant fine powder. The *Lyctus* tunnels in addition are just the colour of the wood in which the tunnels are made. They are not stained in any way.

The most important feature of both Pinhole and Shothole borers from the point of view of the timber importer is that when the moisture content of the timber drops to the extent that the Ambrosia fungi cannot thrive, the boring larvae die. Thus in the dry timber there are no living Pinhole or Shothole larvae and adult beetles are quite unable to infest such timber.

The distinction between Pinhole and Shothole is merely one of size. Pinhole borer is the name given to those species whose galleries seldom exceed a diameter of 1·5 mm. The tunnels of the Shothole borer, on the other hand, reach a diameter of 3 mm, which is approximately the size of lead shot. Incidentally, another important characteristic possessed both by Pinhole and Shothole is the straightness of the tunnels, which is quite distinct from the meanderings of either *Anobium* or *Lyctus* galleries.

8 Wood wasps and marine borers

Wood wasps SIRICIDAE

Wood wasps are fairly large insects classified in the insect order HYMEN-
OPTERA – the ants, bees and wasps. Insects in this order have four wings,
the hinder pair smaller than the fore pair, and the wings of each side
interlock by means of a row of hooklets. The saw-flies, Wood wasps, and
the Honey bee are vegetable feeders, but the vast majority of species in
this very large order are carnivorous, being mostly predacious or parasitic on other insects. Although a number of Wood wasp species have
been imported in timber from various parts of the world, the species
most usually found in Britain are as follows:

Urocerus gigas. The female is a large black-and-yellow insect, very much
like a hornet in general appearance, and although it possesses a large
sting-like projection (the egg-laying apparatus) at the tail end, in fact it
is harmless. The abdomen of the male is brownish-orange and does not
possess the sharply-pointed egg-laying tube. Although usually large in
size, diminutive examples also occur, so that size alone is unreliable for
identification.

Sirex juvencus, *Sirex noctilio* and *Sirex cyaneus* are three species which,
being very similar in general appearance, require a trained entomologist
to identify them. The interested reader is referred to more specialist
works. The females of these species are steely-blue in colour and the
egg-laying tube is prominent (Fig. 43). The abdomen of the male is
brownish-orange with some black at each end. During warm sunny days
the female Wood wasps bore holes in the bark of pine, larch, fir and
spruce trees which are lacking in vigour. The boring is carried out with
the egg-laying tube which is exceedingly tough and furnished with cutting teeth. A few eggs are laid at the bottom of each shaft. A white
secretion lubricates the walls of the egg tubes to facilitate the passage of
the long thin eggs. By means of this secretion the wood is also inoculated
with a fungus carried by the female in special glands. The fungus is the
species *Stereum sanguinolentum* which causes a white-pocket rot in
many species of coniferous trees. The *Sirex* Wood wasps cannot exist
without it, as the hyphae of the fungus grow rapidly preceding the

Fig. 43. A common Wood wasp, *Sirex noctilio* (F.), female. Length about 30 mm.

boring of the *Sirex* larva and play some part in its nutrition. The larvae of *Sirex* Wood wasps are whitish, cylindrical, and have three pairs of very blunt projections underneath the thorax which serve as legs. The larvae, however, are easily identified by the terminal spine at the end of the abdomen, which is of characteristic shape and projects upwards. The length of the life-cycle out of doors varies between two and three years, but indoors this may be prolonged for several years. Softwood timber infested with a few larvae of *Sirex* Wood wasps is not uncommonly used in building construction (Fig. 44), and eventually the adult wasps emerge, which, because of their rather fearsome aspect, usually cause some alarm. They do not, however, re-infest dry wood inside buildings. A point of interest in the biology of *Sirex* concerns their parasitism by two remarkable insects, also in the HYMENOPTERA. The female of *Rhyssa persuasoria* is equal in size to *Sirex* but more slender, and possesses a very long, toothed egg-laying tube. She endeavours to locate the *Sirex* larva in the trunk and then bores down to it and deposits an egg on it. *Rhyssa* passes through its life-cycle in one year. The second parasite is *Ibalia leucospoides*, and in this case the female searches for and eventually finds the egg-shaft made by the *Sirex* female. She then inserts the egg-laying tube down it and lays an egg in every *Sirex* egg she can reach.

Fig. 44. Damage by larvae of *Sirex*. Note tightly packed frass in the tunnels.

The *Ibalia* larva passes the whole of its life, which is about two years, inside the *Sirex* larva.

Wood wasps emerging from timber in buildings are not a serious problem. They are, however, of importance to manufacturers of goods being sent to Australia on account of official regulations which forbid the importation of softwood from countries where these insects exist unless the timber has been fumigated or otherwise treated to terminate possible infestations. As packing cases are made of softwood the problem can be a large one in many industries.

Marine borers

Although, strictly speaking, marine borers are not woodworm, yet it was thought advisable to give some account of them here since they are of some importance in timber preservation.

When timber is immersed in sea water, it is not long before certain organisms effect a lodgment on it, and, if the timber has not been protected mechanically or chemically, boring commences which, if allowed to go unchecked, will destroy it. Whereas, on land, in temperate countries, by far the most important wood-boring animals are the immature stages of beetles and in tropical countries termites (both insect groups), in sea water insects are absent. Two entirely different orders of animals, however, take their place as far as wood-boring is concerned – the MOLLUSCA and the CRUSTACEA.

Fig. 45. Shipworm—*Teredo*. A wood-boring member of the MOLLUSCA. In proportion to the size of the body the shells are much reduced. It occupies the whole length of the burrow.

Shipworm   *Teredo navale*

Teredo is a bivalve mollusc (as is an oyster or mussel), and therefore is not strictly a worm as is implied by its common name, shipworm. The eggs are released haphazardly into the water in very large numbers and from them ciliated larvae are produced which are able to propel themselves through the water. After a time two microscopic shells are produced together with a 'foot'. Its size at this stage is only 0·25 mm in length and when the larva ultimately reaches a wooden surface it attaches itself in position by a thread known as a byssus which is secreted from a gland in the 'foot'. The tiny shell valves are then brought into action by being made to rock backwards and forwards, and by this means the larva quickly burrows beneath the surface of the wood and never returns to the open water again. It continues to bore by the rocking action of the valves, the latter being pressed against the wood by the 'foot'. As the burrow lengthens, so the body of the *Teredo* elongates (Fig. 45) so that the length of the *Teredo* always equals the length of the burrow. As it grows, the *Teredo* secretes a chalky lining to its burrow which is characteristic (Fig. 46). Near the minute opening are two small siphon tubes, through one of which (inhalent) the *Teredo* draws in sea water which is filtered of its small organisms before being passed out through the exhalent siphon. A pair of 'pallets' are able to block up the opening after the siphons have been withdrawn.

When fully grown the *Teredo* can reach a length of more than a metre and a diameter of 25 mm, but the usual size is much less than this. Size mainly depends on the number of *Teredo* present in the wood. They never share a burrow – each being quite distinct although the chalky lining may be the only walls at some points between their bodies. Thus, if only a few *Teredo* are present, they will grow to a large size, but if many are present, they will not grow so large.

It is quite common for hardwood logs imported into this country to show superficial *Teredo* borings. Storage and collection of the logs often takes place in harbours infested by *Teredo*, and during the month or so of floating in the sea the *Teredo* larvae settle on them and start to bore. *Teredo* is able to survive fairly short periods of drying-out of its timber

Fig. 46. Damage by the Shipworm, *Teredo*. The white, chalky lining to the burrows can be seen clearly.

– ten days to a fortnight or so – a factor which has to be reckoned with in dry-docking of wooden-hulled ships. It can also keep alive for a considerable time if the salinity of the water is reduced somewhat. The average salinity of the sea is 35 parts per 1 000, and down to salinity of 10 parts per 1 000 it can burrow and reproduce normally; if, however, the salinity is reduced to 5 parts per 1 000 it can only just exist, and below this salt concentration it will die within a few weeks. This information is important to engineers of harbour works in estuaries where the salinity fluctuates with the volume of fresh water brought down by the river. Temperature also affects *Teredo*. The slightly higher temperature of the sea water in a harbour where there is an effluent from a generating station is said to cause increased *Teredo* activity. *Teredo* activity is much more pronounced in tropical waters.

The Gribble   *Limnoria lignorum*
This wood-destroyer is a crustacean and is classified in the group ISO-PODA, to which the common woodlouse belongs, and indeed, it has the same number of walking legs – seven pairs – as this latter animal. The Gribble is never much more than 5 mm long (Fig. 47) and its tunnel is usually not longer than 50 mm. The burrowing Gribble keeps close to the surface of the wood, entering at a small angle, and here and there small lateral galleries connect with the outer surfaces – these are the

Fig. 47. The Gribble (*Limnoria lignorum*) top view. This is an isopod crustacean in the same group to which the Woodlouse belongs. Length: 5 mm.

respiratory pits or 'manholes' (Fig. 48). By means of these the Gribble ensures a plentiful supply of fresh oxygenated sea water. Each gallery houses a male and a female Gribble, but it is the female which does the burrowing, the male usually being found some little way behind her. When disturbed the male is easily extracted from the burrow, but not so the female. She flattens herself against the end of the burrow and with the sharp claws on her legs is well able to resist removal. Usually no more than twelve larvae are produced at a time, and when these leave the parental brood-pouch they start burrowing side tunnels from that of the parents. Later, when fully-grown, they may leave the burrow and, swimming through water, find other timber to infest.

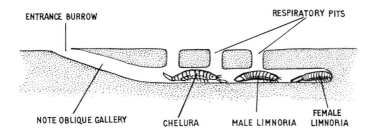

Fig. 48. Gribble (*Limnoria lignorum*) and *Chelura terebrans* shown diagrammatically in their burrows.

Fig. 49. Piece of softwood heavily 'gribbled'.

Gribble-infested wood is decayed in stages. The infestation is always confined to the outer skin of wood, but when this has been completely riddled with the Gribble burrows, wave action washes it off so that a fresh surface of wood is presented to the Gribble to destroy. The appearance of Gribble-infested wood is quite characteristic and is very easy to indentify (Fig. 49). Gribble requires water of high salinity, so that it is not found in estuarine waters.

Fig. 50. *Chelura terebrans* – side view. An amphipod crustacean usually found in the Gribble burrows. Length: 5 mm.

When examining Gribble-infested wood another small animal is sure to be extracted from the Gribble burrow. This is another crustacean, a member of the group AMPHIPODA, to which the sand-hopper and the freshwater shrimp belong. Its scientific name is *Chelura terebrans* (Fig. 50) and it is thought that some symbiotic relationship exists between it and *Limnoria*, i.e. that they enjoy mutual benefits from living in association with each other.

9 Identification of damage to timber

When timber is being examined for evidence of insect damage, the adult insects are likely to be present only during the limited flight season of the species concerned. On the other hand, the larvae and pupae are inside the wood, only to be found and identified if the wood is carefully split up and thus destroyed. This is seldom possible in practice.

In the following pages, therefore, the identification of the wood-boring insect is considered in its absence, the characters which are given being those of the damaged wood, the type of wood likely to be attacked, the nature of the tunnels and flight holes and the appearance of the faecal matter and rejected wood particles (Fig. 51).

Fig. 51. Larval frass of: (a) Furniture beetle, (b) Death watch, (c) *Ernobius*, (d) Weevil, (e) Powder post, (f) House longhorn. All magnified ten times.

	TIMBER				LARVAL GALLERY		FRASS		FLIGHT HOLE	
Species	*Species*	*Sap or Heart*	*Fungal decay*	*Shape of transverse section*	*Straight or Meandering*	*Stained or Clean*	*Quantity*	*Description*	*Shape*	*Size*
Anobium punctatum	Wide range of softwoods and hardwoods	Mostly S but sometimes H	Not present or very slight	Circular	M	Clean	Moderate	Cigar-shaped pellets. Gritty feel	Circular	About 1·5 mm
Xestobium rufovillosum	Hardwoods oak, chestnut; rarely softwoods if adjacent	S and H	Always present	Circular	M	Clean	Moderate	Bun-shaped pellets	Circular	About 2·5 mm
Ernobius mollis	Spruce, pine, larch, fir	Outer sapwood where bark adhering	Absent	Roughly circular, sometimes oval	M	Clean	Moderate	Bun-shaped pellets present of two colours according to whether bark or sapwood has been eaten	Circular	Up to 2·5 mm in bark
Ptilinus pectinicornis	Beech, elder maple, oak, alder, plane, sycamore, elm, hornbeam, poplar	Sapwood	Absent	Circular	M	Clean	Moderate	Fine particles only. Silky	Circular	About 1·5 mm

Euophryum confine Pentarthrum huttoni	Softwoods and hardwoods	S and H	Present; often wet rot	Indistinctly round	M	—	Small	Fine in texture round granules	Narrowly oval; ragged margin	About 1·1 mm
Lyctus sp.	Wide pored hardwoods eg: oak, ash	Starch rich sap	Absent	Circular	M	Clean	Copious, sometimes bursting undamaged outer skin	Fine particles silky feel	Circular	About 1·4 mm
Hylotrupes bajulus	Softwoods but recorded in some hardwoods such as oak, alder, poplar	Sap first but heart at late stage	Absent	Oval	Straight in early infestations later meandering	Clean 'ripple' teeth marks	Copious, often bursting veneer of unattacked surface wood	Fine particles (rejected) with well formed cylindrical faecal pellets	Oval	Up to 6 mm long axis, but flight holes made by males often very small
Phymatodes testaceus	Usually oak but chestnut and ash recorded	Outer sap	Absent	Flattened oval and broad channels	Mostly with grain	Clean	Large quantities sufficient to loosen bark	Roughly cylindrical with one or two constrictions; irregular ends	Oval	About 6 mm long axis
Clytus arietis	Usually beech and oak but fruitwoods etc., not uncommon	Under bark until half grown then sap	Absent	Oval	Meandering	Clean	Moderate	Fine particles and loose aggregates	Flattened oval	About 5 mm long axis

Species	TIMBER		LARVAL GALLERY			FRASS		FLIGHT HOLE	
	Sap or Heart	Fungal decay	Shape of transverse section	Straight or Meandering	Stained or Clean	Quantity	Description	Shape	Size
Nacerdes melanura	S and H	Always present	Large irregular channels	Usually with grain	Clean	Moderate	Soft and wet with fibrous shavings	Ragged oval	About 5 mm long axis
Buprestis aurulenta	S and H	Absent	Flat oval	Mostly with grain	Clean	Tightly packed in arc-like pattern	Irregular spherical	Oval	Up to 5 mm long axis
SIRICIDAE *Wood wasps*	S and H	Present but not usually detectable	Circular	Curved	Clean	Gallery tightly packed	Coarse	Circular	Up to 8 mm
SCOLYTIDAE PLATYPODIDAE *Ambrosia beetles*	S and H	Presence shown by staining	Circular	Very straight	Stained	Nil	—	Absent from sawn timber	—

Special note: the description of the wood damage caused by SIRICIDAE is equally applicable to the melandryid beetle *Serropalpus barbatus* which is common in Europe but only occasionally introduced into Britain.

10 Control measures

The subject of woodworm extermination and prevention is a large and complicated one, and is beset with many problems and side issues. It is realized that no simple answer to this question presents itself. In the first place, the term 'woodworm' includes a number of biologically distinct insects, and the differing life-cycles, types of timber attacked, differing ages of timber attacked and different parts of the wood tissue involved make identification of the species of the wood-boring insect a first essential. Knowledge of the biology of the insect concerned is therefore required.

The measures to be taken against any particular infesting insects must be closely related to this information. The identification and biology of those wood-boring insects which occur in the timber-yard, the cabinet-making factory, in furniture and structural timber in our homes – new and old – has now been discussed. This chapter outlines the measures to be taken for prevention and destruction of these pests.

TREATMENT OF FURNITURE

Almost all the woodworm damage to furniture is caused by the two species Common furniture beetle (*Anobium punctatum*), and the Powder post beetle (*Lyctus* species). The latter beetle infests 'new' furniture made from one or more of the susceptible wide-pored hardwoods, such as oak, chestnut, ash, abura, khaya and limba, and there are, of course, many others. Only the sapwood is attacked, and only when the starch content is above a certain minimum figure. This has already been explained in greater detail. 'Live' infestation of *Lyctus* will normally be found only in furniture manufactured within the last ten years.

Anobium will be found mainly in older furniture. An infestation may not be expected in the solid wood of oak furniture until some 60 years have elapsed from the date of manufacture. Birch and alder plywood will often be found to be attacked, however, within three or four years of its manufacture. This information means that when a furniture manufacturer receives some woodworm-infested articles returned by the distributor, the insect is invariably *Lyctus*. Only one case to the contrary is known to the writer and this was where some sheets of birch plywood had been used which were obviously from a parcel which had been in stock a long time and located during a plywood shortage.

Some differences in procedure in dealing with woodworm will, of course, be usual according to whether this is carried out by a furniture manufacturer or whether it is being dealt with in the home.

Home remedies

Almost every column written about woodworm in magazines mentions one or two 'home remedies'. Quite a common one is paraffin or turpentine which may or may not be applied to the holes with a fountain-pen filler. There is no doubt that if the sapwood of softwood is being treated, and provided also that some of the woodworm larvae are near the surface of the wood, paraffin would kill some of them, but such materials would by no means exterminate all of them and the effect would be temporary only. Both paraffin and turpentine are very volatile and, therefore, dry out within a comparatively short space of time. Shortly after such treatment a female beetle alighting on the wood may lay another batch of eggs, and as the paraffin or turpentine would have vaporized, re-infestation would commence at once.

Oil-based woodworm fluids

An advantage held by fluids with a non-volatile base, as against those with a volatile base, is that as soon as the latter dry out or lose their solvent to the atmosphere, penetration into the wood obviously ceases, leaving the insecticidal materials at the level of penetration reached when this occurs.

On the other hand, those fluids containing an appreciable percentage of non-volatile mineral oils exhibit the phenomenon of 'creeping' or continuous penetration of the wood. This has been demonstrated as occurring even after several years have elapsed from the time of initial superficial brush treatment. However, the deeper the insecticidal chemicals penetrate, the less concentrated they become, and only reputable brands with appropriate chemical loadings to allow for this factor should be used.

When a piece of furniture is found to be attacked by woodworm, treatment for its extermination should be carried out without loss of time. It will be appreciated that the occurrence of flight holes will signify that at least one generation of *Anobium* or *Lyctus* has completed its life-cycle and it would be almost certain that one or more overlapping generations of larvae would still be present in the wood.

The reader will also now appreciate that infested furniture taken into an uninfested house places the house at risk, and new furniture placed in an infested house is also undesirable.

The view put forward, from time to time, that insecticidal treatment against woodworm should be undertaken only at the time of emergence of adult beetles from the wood, is mistaken. The semi-permanent effect of the proprietary oil-soluble woodworm fluids makes their effect independent of season of application. The simple treatment outlined below should be given to any furniture immediately it is suspected of harbouring woodworm, or if it is proposed to prevent possible woodworm attack in the future.

Overall brush treatment

Remove all dust and dirt; then, in the case of upholstered articles it is desirable to remove any leather or textile to prevent staining.* It is then best, after removing any drawers or shelves, to turn the article upside down and thus obtain more or less, the Furniture beetle's-eye view. Apply one or more liberal coats of the proprietary woodworm fluid which has been chosen, according to the manufacturer's instructions which will be found printed on the tin, and if there is any special coverage, such as one gallon to so many square feet, make sure that this amount at least is applied. Work the fluid into all the cracks, crevices and rough surfaces with the brush. The bottom surfaces of legs should not be forgotten, as perhaps, in furniture, this is the most common part for the beetle to attack first. It would be the first part to be reached by a female beetle walking over the floor in search of an egg-laying site and, of course, helped by the fact that the bottom surfaces of the legs are almost invariably left in a very rough condition.

An important distinction between the various proprietary fluids must now be understood. Some materials are not suitable for treatment of polished surfaces, whilst others do not harm such surfaces. Fluids coming within the former category will have this fact brought to notice on the manufacturer's 'directions for use'. If the fluid which is being used belongs to the latter group it should be liberally brushed over all the polished surfaces, after the unpolished surfaces have been treated. The drawers and shelves which have temporarily been removed, should be treated in the same way.

Injection

Almost all furniture is made of hardwoods which in the main resist the penetration of fluids to a greater degree than softwoods. It is, therefore, always advisable to inject the fluid through some of the flight holes and thereby also obtain an immediate kill of larvae which may be boring deep in the wood. This is, of course, in addition to the overall brush coats. Several different types of injector exist, but perhaps the most successful model uses a plastic container (constructed of polythene) as the bellows (Fig. 52). It is possible to remove the down-tube in some injectors, thus allowing them to be used upside down. Woodworm fluid in an aerosol is also available (Fig. 53).

---

* If this is impracticable then ensure that no fluid is applied nearer than 25 mm to any such material.

Fig. 52. Perhaps the simplest and at the same time the most successful type of injector for use against woodworm in furniture. The plastic container allows the operator to obtain an excellent pressure. Note the all-important nozzle washer which enables a tight closure to be made without marking the surface of the polished furniture.

Fig. 53. An alternative injector to the squeeze bottle shown in Fig. 52 is this aerosol container. Again, a special injector nozzle is provided to reduce the amount of fluid escaping.

The mode of action is as follows: select some of the larger flight holes and insert the nozzle of the injector into one of them, pressing the rubber washer against the surface of the wood; the subsequent pumping of the fluid into the wood will ensure full pressure, forcing the fluid along the intercommunicating galleries, often deep into the wood. The rubber nozzle washer is of the utmost importance, since it prevents fluid squirting back into the eyes or onto furnishings some distance away, and ensures that most of the pressure is used to force the fluid deep into the wood. This is the reason why fountain-pen fillers, oil cans, hypodermic syringes, etc., lacking this part are not nearly so efficient. Spraying devices introduced into flight holes are sometimes used, but with little success because the fluid being sprayed is deposited on the walls of the gallery at its beginning only and the air merely blows out bore-dust. The aerosol type injectors are however an exception to this. The positive action of injectors is often demonstrated in dramatic fashion. Fluid will often emerge from a flight hole some distance, as much as a foot away, from where it is being pumped in. This is, of course, because of the way in which the tunnels intercommunicate within the wood. It follows that the worse the infestation, that is, the more intercommunicating tunnels there are, the easier it is to introduce the woodworm-killing fluids right inside the wood if a good injector is used. Care should be taken not to exert too great a pressure which might cause splashing or deflection backwards into the eyes.

Filling
If any fluid remains unabsorbed on the surface of the wood after a few days, it should be removed with a dry cloth, and it is probably a good idea to fill in the flight holes, if they are not too numerous, with a filling material. For light woods molten beeswax can be pressed into the hole with a knife; this is, of course, the good old standard way, but with darker furniture it is a good plan to fill in with one of the plasters such as 'Polyfilla'. These are white, but have the property of absorbing colour and can be tinted to give the exact match (an 'Oxo' cube is excellent for the purpose) and be polished afterwards.

The filling of flight holes serves the useful purpose of indicating fresh ones, should they occur.

If a chair or a pew has been treated it is often advisable to seal the woodworm-killing fluid with a wood sealer. This will prevent clothes from spoiling if the chair is to be used immediately.

Period of immunity
It is often asked how long furniture or woodwork generally, treated with one of the modern oil-soluble wood preservatives, will be kept free from woodworm attack. The writer, of course, is not able to speak for all types of fluid, so that it is difficult to give a general answer, but having a great deal of experience with one particular type of fluid, it can be said

that in this case, at a very conservative estimate, the period is from twenty to thirty years; indeed there is strong evidence to suggest that the period may be much longer.

Wetting action of fluid
The fluids which have been mentioned above are almost all light-brown to colourless and have no pigment or dye in their formulation. In order to be effective, however, the surface of the wood must be wetted by the fluid, and as the light colour of many woods is due to air in the surface fibres, the wetting action will have the effect of toning the wood a slightly darker colour. As the fluid is absorbed, however, by the drier wood underneath, the original colour will return. This means that it sometimes happens, in imperfectly polished wood, that some part of the surface is not wetted, and this gives a streakiness which may persist for some months. The extermination of the woodworm is, however, immeasurably more important than a temporary colour blemish.

Due care should be taken to prevent splashing the fluid on plaster work.

Fumigation
This process consists of placing the infested article in a container or chamber which can be made gas- and pressure-tight and then introducing the fumigant by means of a valve. The fumigant is usually in the form of a gas, examples being hydrogen cyanide and methyl bromide. A specified time is allowed to elapse before the gas is drawn off or allowed to escape into the atmosphere. A further minimum period of ventilation before the articles are handled and subsequently delivered back to their owners is allowed as a safety precaution. Both gases given as examples are very poisonous and should be used only by skilled and experienced operators. Fumigation is a method of insect control often employed against, for example, stored-food insect pests, and its employment is sometimes recommended for use against woodworm. It must be remembered, however, that fumigation by itself acts in a temporary manner only. As soon as the fumigant is dispersed (within a few hours of the treatment) there is nothing to prevent re-infestation, so that fumigation is usually employed in conjunction with the use of long-acting oil-soluble woodworm-killing fluids or special insecticidal polishes. This combination of treatment is most useful on items which for various reasons may not be treated with oily fluids or where upholstered furniture or large quantities of furniture are to be treated. Obviously, by the nature of the fumigation process it must be carried out by firms of specialists licensed for the purpose.

Heat treatment
Placing furniture in an oven or kiln has been suggested for use where woodworm occurs in polished furniture. It is stated that temperatures of

55°C (130°F) and relative humidities of 80% do not appreciably affect surfaces finished with cellulose paint, french polish or turpentine varnish.

In addition, several types of short-wave radiation have, from time to time, been suggested as a means of tackling the woodworm problem but there seems to have been little commercial interest taken in these methods, as indeed in heat treatment. The reason for this appears to be the comparative ease of application and the extraordinarily high degree of efficiency and long-lasting powers of the oil-soluble woodworm-killing fluids, whereas heat treatment and short-wave radiation would have a temporary effect only.

TREATMENT OF STRUCTURAL TIMBER

By far the most important part of the woodworm problem in the United Kingdom concerns the widespread occurrence of *Anobium punctatum* in the structural timbers of buildings. As previously mentioned, this almost certainly concerns more than half the total number of dwellings, numerically something like 11 000 000, besides an incalculable number of farm buildings, sheds, industrial properties and other buildings. Beside this, the problem as concerns furniture pales almost into insignificance.

Before carrying out any work against woodworm in the structural timbers of a building, an accurate survey must be undertaken to ascertain the nature and extent of the infestation and its distribution. Access should be made into the roof space, the 'cupboard under the stairs', basements and cellars. Indeed, *all* woodwork must be thoroughly examined and reported on.

The surveying operation alone often entails much physical discomfort and painstaking work, and whilst some householders may be capable of carrying it out competently the great majority consider that this task and the subsequent control work must be the responsibility of a woodworm-servicing company.

The *in situ* treatment of building timbers against infestations by *Anobium punctatum* by servicing companies has become a highly specialized industry (Fig. 54) and although the householder by this means is able to place the responsibility for carrying out the work on other shoulders it would be very wise to make himself familiar with the general principles of the work.

Making the decision

As a householder, the first stage in woodworm control is to make the decision to have it carried out. The question often arises, when making this decision, whether the woodworm control treatment is best carried out during a particular season, such as just before emergence of the adults. Such a season for treatment, say in early June, would be valid

Fig. 54. A full kit of special tools, labour-saving devices and protective
clothing used by a servicing team of woodworm exterminators.

only for those beetles emerging in that season, whereas there will remain
in the wood larvae which will not be emerging till the following year and
the year after that. In fact, the long-lasting properties of woodworm-
killing fluids make their application independent of season. The best
time, therefore, to have treatment carried out is as soon as possible after
the decision has been made, as is mutually convenient to the householder
and the treatment firm.

There are other aspects concerning the woodworm control decision
which should be mentioned here. Why is the decision made? Why do
two or three householders out of four, when confronted with the decision, decide to get woodworm control under way as soon as possible?
There are several reasons. Perhaps the most important relates to the
large capital investment represented by the purchase of a house. This is
usually the most important capital investment ever made by the average
man, so that when he realizes that some part of the structure is progressively deteriorating he feels compelled to do something about it.

In addition to this reason for the decision, however, over the past few
years there has occurred a very strong feeling that insect pests of any
description in the home should not be tolerated. Complete freedom
from insect pests is, indeed, one of the criteria of high living standards.

In a high proportion of cases the presence of woodworm only comes
to light during a change of house-ownership. It commonly occurs that a
surveyor employed by a prospective purchaser establishes the presence

of the pest, for example in the roof void, and then recommends his client to make an offer much less than the price being asked. On the other hand, the vendor who is already aware of the presence of woodworm in his house often, wisely, gets the woodworm trouble cleared out of the way before putting his house on the market. Wisely, because the sum offered by a prospective purchaser whose surveyor has found a woodworm infestation is invariably considerably less than the asking price less the cost of woodworm treatment. Moreover, the presence of woodworm in the home, shown by the groups of little holes appearing in well-loved furniture or in the structure of the home, the little brown beetles and the piles of frass, cause worry and anxiety, and in extreme cases when other worries or anxieties are present as well, can give rise to mental ill-health of considerable severity. All those who have experience of working in woodworm control in direct contact with the public are able to give many instances of this.

The survey
The first essential part of a woodworm control operation is the survey. This must be conducted by a thoroughly trained* timber infestation surveyor who not only can identify with accuracy all the different species of wood-boring beetles and wood-rotting fungi likely to be found in buildings, but who can estimate the distribution and extent of the damage likely to have been caused. The degree of attack by any of the insect and fungal decay factors must be assessed, and the best means of control specified. The timber infestation surveyor, unless specially instructed to do so, will not normally inspect outbuildings, gates and fences outside the building, nor furniture inside. An underfloor inspection is very necessary as, although floor-boards often show little or no woodworm attack on the top surface, the underside and the floor joists may often show considerable infestation (Fig. 13). The timber infestation surveyor, therefore, will require to lift a floorboard or two and use his torch in order to make an underfloor inspection.

Much can be done by the householder to help the timber infestation surveyor. Fitted carpets and other floor coverings can be loosened, and where possible rolled back. Bulky stored articles should be moved away from timbers, especially in the roof, and more especially from around the trap door. It is equally important to remove stored articles from the vulnerable 'cupboard under the stairs'. It is remarkable how frequently this space shows the first signs of Furniture beetle attack. In the case of large buildings adequate ladders should be available for inspection of otherwise inaccessible woodwork. It is surely obvious that it is in the best interests of the householder to co-operate with the timber infestation surveyor and to grant him every facility in carrying out his task. The

---

* In 1980, examinations conducted by the Institute of Wood Science were instigated and this welcome innovation should improve technical standards.

surveyor divides the building into its separate building units; e.g. the roof voids, the various floors, the staircase and the joinery and these are separately examined and reported upon, because a rather different specification is applicable to each.

The treatment method
The basic treatment for wood-boring beetles in structural and joinery timbers has a common application to most of the insects detailed in previous chapters. Some exceptions are mentioned later. Methods used are outlined below: more complete specifications are usually given to clients by the timber infestation surveyor with his report.

Preparatory work
All exposed timbers must be cleaned prior to treatment. Usually light brushing is all that is required, particularly to remove spiders' webs, but rarely it may be necessary to remove dust and debris with an industrial vacuum cleaner. Also, some of the electrical installations may require protection with polythene sheeting and the mains should be disconnected (see Health and Safety, p. 94).

All this preparatory work may be undertaken by the servicing company, or alternatively the householder may be responsible for this work.

Application methods
The application of the chosen woodworm fluid by brush is the method used by most D-I-Y householders. It is a time-consuming job and is therefore only used by professional servicing companies where special care in application is required.

Faster application is achieved by spraying (Fig. 55), but this method calls for specialized equipment not readily available to the householder. More even application of the woodworm fluid is obtained by spraying, and this is the only method suitable for work in restricted areas (e.g. eaves) in roof voids where it is often impossible to reach all surfaces with a brush.

Where joinery timber is being treated, a surface application should be supported by injection of fluid into flight holes. This can be done with small hand injectors (Figs. 52 and 53), but the professional's pump-operated injector will give greater pressure and more speed.

Some treatments are now being made with an 'embodied mayonnaise' type preservative. The insecticide is incorporated into a soft paste which is applied in strips to the timber. Gradually, the material creeps out over the surface and the insecticide soaks in.

Where to treat
All exposed timbers should be carefully covered with the fluid at the rate specified on the can. Always read the label fully before starting work.

Fig. 55. Spraying floor-boards and joists with woodworm fluid. Note the use of full protective clothing for this work.

Treatment in roof voids is usually straightforward except where access is restricted (see above). All surfaces should be treated – remember that woodboring larvae may be tunnelling in timber that *appears* to be sound. Special attention should be paid to end grain which provides an ideal egg-laying site for Common furniture beetle. Special safety precautions should be followed (see Health and Safety, p. 94).

In rooms, sufficient floor-boards should be lifted to allow both sides to be treated, together with the exposed joists and framing timbers. Staircases require special attention, since it is essential to reach the underside of all timbers. This may not be possible without structural upheaval: holes then have to be drilled and special nozzles are required to spray the back surfaces.

Types of preservatives

So far in this chapter, we have mentioned two types of preservative – oil-based fluids and embodied mayonnaise. It is not proposed to go into the complexities of formulation detail except to state two positive points, firstly that consideration of a woodworm-killing fluid to be widely used in all possible circumstances must entail a study of the human toxicity factor. The effect of accidental drinking by children and possible criminal use must be established accurately and guarded against by the possible use of an emetic 'built-in', or by other means. Cheap formulae have sometimes been published in professional journals which would not have stood the test of this elementary humane principle.

The second point concerns the 'effective dose' i.e. the woodworm-killing fluid must be at such a concentration that when a certain amount is supplied evenly over a given area it will be effective in 100 per cent of cases. Preferably from the economic standpoint such a concentration should be capable of application at one coat without 'run off'. This has been a very difficult exercise in view of all other essentials of formulation, but was achieved some years ago.

The final result of this is that provided certain rules of formulation are followed in the make-up of the woodworm-killing fluid and certain rules of fluid application followed rigidly, the elimination of *Anobium punctatum* from wood which it is infesting is virtually certain. What is more, the wood can be made impervious to future attack for many years.

Oil-based fluids are most often used, but other types are available, and also there is sometimes a choice of active ingredient. The high cost of petroleum products for the oil-based sprays has led to research into 'water-based' fluids, but so far few are available commercially (and none are on retail sale). The main difficulty to be overcome is the fact that most conventional insecticides are not soluble in water and it is still necessary to dissolve the insecticide in a light oil first. Thus, the 'water-based' fluids in use today are really 'emulsions' with the insecticide dissolved in minute droplets of oil suspended in the water. The ready-to-use fluid still contains about 50% oil.

Oil-based fluids usually contain the insecticides lindane (γ-HCH, BHC) or dieldrin (although the latter is no longer used by some companies). Both insecticides are extremely long-lived and if properly formulated and applied, enable specialist companies to guarantee their treatment for 30 years (see Chapter 11). Other insecticides used for woodworm control include some of the modern synthetic pyrethroids, but these have yet to be proved in practice.

Smoke generators and dichlorvos strips

A large proportion of *Anobium punctatum* infestations occur in the relatively confined areas of roof spaces. In the past, lindane smoke generators have been used in an attempt to kill invading beetles and those which emerge from infested timber already present. This method is not particularly effective, since the small particles of insecticide settle out on only horizontal or sloping, upward-facing surfaces. This leaves most roofing timbers completely unprotected. Also, of course, the insecticide does not penetrate the timber to kill woodworm larvae.

More recently, dichlorvos impregnated resin strips have been advocated. In unlined roof spaces (typical of pre-1939 houses) the excessive ventilation reduces the amount of insecticide in the air and few beetles are killed. Conditions in modern, lined roof spaces are such that most adult beetles are killed before they can complete their exit holes. Any

eggs laid on the surface are killed before hatching, but those laid deep
in old flight holes will produce viable larvae. Larvae already established
in the wood are not affected by the insecticidal vapour.

Thus, dichlorvos strips cannot replace liquid insecticides formulated
to kill the larvae in wood. Used every year from June to August, an
existing infestation may gradually be eliminated, but the method would
be quite expensive when extended over 20 or 30 years. Also, no written
guarantee can be obtained and there is nothing to show to prospective
purchasers when the house is to be sold.

On the other hand, when insecticidal fluids are applied by skilled
operators, all existing larvae are killed, and any larvae hatching from
eggs laid subsequently will also be killed. When completed, the owner
obtains a written guarantee indicating the extent of the treatment work.
The purchaser then knows exactly, by specification, which timbers have
been treated, and the documentation retains its value against the rising
price of property.

Fumigation of houses

It is rarely necessary to fumigate buildings to control wood-boring beetles, although the work is sometimes carried out for House longhorn in
Scandinavia and for Common furniture beetle in New Zealand. Termites
are also treated in this way, see Chapter 12. The method is extremely
expensive since the building must first be made gas-tight by sealing or
covering with gas-proof sheets. The gases used, methyl bromide and
phosphine, are extremely poisonous and can only be applied by specialists.

The most important aspect to bear in mind, however, is that fumigation kills only those insects which are in the building timbers at the time.
The premises are open to re-infestation as soon as the gas has dispersed.

Carpentry work

It will be seen from the treatment methods above that a considerable
amount of what can be called carpentry work is entailed – such as the
lifting and subsequent relaying of floor-boards. So in comparing esti-
mates submitted by two or more servicing companies, the appropriate
specification also should be compared, as some servicing companies
undertake no carpentry work; leaving this to the householder's builder.
With other companies the estimate is inclusive of all such work. Like
should be carefully compared with like. It has always been the opinion
of the author that woodworm control is a specialist business and full
responsibility for it should be borne by the treatment firm, not hived off
on to a building company.

In addition to the relaying of original timbers like floor-boards, it may
be necessary to discard heavily infested pieces and to fit new timber.
With the use of smaller dimensioned timber in modern houses, the possibility of it being weakened by the activity of wood-borers becomes

even more likely. From an economic point of view, then, it makes good sense to treat such timber long before replacement becomes necessary.

Pre-treatment
All such remedial problems could be avoided by the inclusion in new houses of timber pre-treated to prevent attack by wood-boring insects (and wood-rotting fungi). Today's primary method, the vacuum/pressure impregnation of timber with a copper/chrome/arsenic mixture (e.g. 'Celcure'), gives a product which should last the length of life of the building itself. The additional cost of pre-treatment is minimal when compared to the total outlay on the whole building. Full details are given in another book in the Rentokil Library, *Industrial Timber Preservation*.

SPECIAL TREATMENT METHODS

The foregoing has referred principally to the treatment of timber infested with Common furniture beetle, *Anobium punctatum*. In general, such methods apply to all other wood-borers, but there are some additional points to make.

Death watch beetle
Because of the association of Death watch with fungally-decayed wood emphasis must be given to the eradication of the environmental conditions giving rise to fungal attack. The application of fungicides is sometimes valuable. Ventilation of the wood must be increased materially, and any sources of excess moisture removed. The insecticidal fluids must be applied at heavy dosage rates and injected into the wood, where possible through old flight holes. Death watch attack often takes place in cavities within large dimensioned timbers, and these must be reached by injection if suspected.

Injection can, of course, only take place where there are flight holes, but a method of introducing the fluid right inside the wood, often carried out with success, is to bore a series of holes as near to the top of the beam as possible, and sloping downwards. These holes are then filled with fluid which is replenished from time to time. Funnels made to fit tightly in these holes by means of corks allow more fluid to be introduced than would otherwise be possible.

Alternatively, plastic injectors containing a non-return valve are now available which are permanently fitted in holes bored in the wood. Fluid is forced through each injector and is retained within the timber. Also available are pressure injectors which deliver a measured dose of fluid to each hole without the need for the non-return valve unit. High pressure is maintained by means of a special expanding nozzle which prevents fluid squirting from the hole.

It is amazing how often a severe attack of Death watch can occur in a church roof without it being observed. Sometimes at the time of emergence the adult beetles are present in thousands; at other times, where there is an equally severe infestation (shown by fresh crops of flight holes) the adult beetles are by no means so apparent. The presence of bats feeding on the beetles has been held to account for this. Again, where there is a severe Death watch attack, the spread of the attack to softwood should be expected, as in these circumstances it is not uncommon. These softwoods also show fungal infection. This is, however, quite a different phenomenon from the apparent infestation of deal boards lying on top of infested oak joists. The Death watch flight holes only represent lines of escape from the joists. The deal boards in this case are usually sound and free from fungal infection. In treating such cases, as well as preservative treatment, the joists should be insulated from damp.

There is a general impression that Death watch extermination is exceptionally difficult. The reason for this impression appears to be inadequate rather than ineffectual treatment. Some timbers are treated that are known to be infested, and perhaps the following year a number of Death watch beetles are found on the floor and the observation is made that the treatment has done no good. A check should be made to find out where the beetles have come from. In the experience of the writer these beetles have often emerged from timber which has not been treated. This brings us back to the golden rule for Furniture beetle extermination, which should, as far as economically possible, be applied to Death watch – treat the whole of the job.

Wood-boring weevils
As with Death watch, timber infested by weevils is usually degraded by fungi and should be replaced or treated with a fungicidal fluid. Then make sure that sources of excess moisture have been removed and the ventilation improved, if necessary.

Powder post beetles
As we have seen, infestations of Powder post beetles usually start in the timber yard, where the adults lay their eggs on freshly felled timber. Thus treatment should largely be directed towards the commercial timber yards. Insecticidal spraying has been practised in Britain for many years and this, together with great improvements in site hygiene – particularly removal of waste timber – has resulted in almost complete freedom from this problem in British timber yards.

Today, almost all *Lyctus*-infested timber is imported in made-up furniture and wood-block flooring. The attacks will only be present in the sapwood sections, but often the infestation is so far advanced by the time it is noticed that replacement is necessary. Treatment of furniture or wood-block floors is often difficult due to the presence of varnish or wax polish and to the binding layer below wood-blocks. Readers are recommended to seek professional advice.

House longhorn beetle

Any sapwood tunnelled to such an extent that it can no longer bear any strain should be cut away. Such sapwood would only soak up large quantities of insecticidal fluid and serve no useful purpose. When all the powdered sapwood has been removed the whole area should be thoroughly cleaned, all dust and debris being removed, preferably by vacuum cleaner. All the woodwork – rafters, ceiling joists, roofing battens, etc. – should be given two or three liberal coats of insecticidal fluid. This should be carried out by brush or coarse-nozzle spray.

Wharf borer

The presence of these beetles in buildings away from water leads to enormous difficulties. Most often the infested timber is buried in the soil, sometimes *under* the building. Thus treating the timber, or even locating it, is frequently impossible.

When timber *is* found, then it is often badly damaged by wood-rotting fungi and by the wharf borers and will need cutting out and replacing. Treatment of adjacent timber and new timber is best made with an oil-based fungicidal fluid, and if possible, the area should be dried out. This will kill both rot and insects. In the case of wooden-hulled boats, a water-soluble fungicide may be added to the bilge water.

Shot-hole and Pin-hole borers

As indicated previously, these insects die out when the timber dries. Thus, no treatment is required.

TREATMENT OF MUSICAL INSTRUMENTS

With the increase in Common furniture beetle it only required the action of the laws of probability for such articles as old violins, which had for two or three hundred years not received a visit from an egg-laying Furniture beetle, suddenly to become the objects for their attention. This is not to say that none of these old instruments ever showed signs of woodworm damage in the past (Fig. 56).

The owners of old violins and other valuable and historic wooden objects are naturally anxious to preserve their property – the Stradivarius violin probably represents wood in its most valuable form – but one is always asked: 'will this insecticide treatment affect the tone in any

Fig. 56. The back of a Nicolaus Amati violin (1653) showing old damage
by Common furniture beetle.

way?' The answer cannot be definite, but it is believed that the tone is
scarcely affected, if at all. Certainly no one who has carried out this
treatment has ever informed the writer that deterioration in tone has
occurred.

The inside of a violin makes an excellent rearing site for Furniture
beetles. I should imagine the female enters the belly of the instrument
through an 'S'-hole and there would find many admirable egg sites. The
application of a fluid to the outside of a fiddle does not present many
difficulties, but there will have to be some ingenuity displayed – with a
camel-hair brush tied on to a piece of wire – before all surfaces of the
inside have had a brush treatment. But take one important precaution
– do not apply the fluid to the pegs, or they will slip.

Pianos constitute another problem. A very large number of pianos
came into this country from Germany at the turn of the century. They
had beech frames, and quite a high proportion of them are now infested
with Furniture beetle. The essential process for extermination remains
the same – a coat of proprietary oil-soluble woodworm fluid applied by
brush to all surfaces and injected here and there where there are flight
holes – say every 10 cm or so.

A piano, again, presents some problems concerning the application of
fluid to the inside, mainly on account of weight – but a few splashes of
fluid on to the wire strings will cause no harm. Incidentally, the same
fluid will prevent rust and keep moths away from the felts.

HEALTH AND SAFETY IN WOODWORM CONTROL

Before using any pesticide in the home, read the instructions on the label on the container and follow them exactly. As with any chemicals, keep woodworm fluids out of reach of children and do not transfer them to unlabelled jars or tins: this particularly applies to bottles which have previously contained drinks. Remember that the health of the young and infirm are at risk if these recommendations are not followed.

The safety precautions described in the next few pages apply to most woodworm fluids: there is even a fire hazard with 'water-based' fluids since they contain sufficient flammable material for precautions to be necessary.

Fire hazard
During spraying and for a short period afterwards flammable vapours will be present which can be ignited. A small spark from a faulty light fitting or a static discharge from a nylon boiler suit is all that is required. In enclosed spaces, the solvent vapour may be present for some time, depending on the degree of ventilation available (unless specialized treatment methods are employed).

Insulation materials
Glass fibre, mineral wool and other insulation materials retain some of the excess fluid which is present during a roof space treatment. However, most of these materials are not absorbent; instead, solvent is held for some time on the surface of the fibres, where it constitutes a fire hazard. All such materials must therefore be removed prior to treatment. They should not be replaced until all the solvent has evaporated: under normal conditions this would be at least fourteen days.

Pipe lagging
Old type hessian pipe lagging *is* absorbent and the fluid will penetrate the fibres. Such lagging can either be removed or covered with polythene sheeting before treatment begins.

Electric lighting
If the electric light bulb or a fluorescent tube is accidentally broken whilst treatment is in progress, a fire may well occur. Therefore, the mains should be disconnected or the relevant fuses removed and a suitable recommended lamp on an extension cable should be plugged into a socket some distance from the work area.

Static electricity
As already mentioned clothes fabricated of man-made fibres, hair combing and all other means of creating a spark by static electricity may cause fires.

Fire extinguishers

Whilst spray treatment is in progress it is advisable to have a fire extinguisher close to hand. The extinguisher should not contain water: dry powder or carbon dioxide types are most suitable.

Rubber-covered electric cables

Fluids used in woodworm treatment can cause rubber to perish and therefore it is necessary to protect cables. This can be achieved by, for example, painting with a wood sealer.

Other action before work commences

Water tanks should be covered with polythene sheeting, on top of which is placed a cotton dust sheet, to prevent contamination of the water by dust and spray.

All fires, naked lights, heating appliances and gas pilot lights in the area of treatment should be extinguished and allowed to cool.

Personal protection

When treating timbers with liquid insecticides (or fungicides), care must be taken to prevent these preservatives coming in contact with the body and it is advisable to avoid inhaling vapour or spray. Trained men from service companies are always well-equipped (Fig. 55) for their own safety (as required by the Health and Safety at Work Act). Householders cannot be too strongly advised to observe the precautions printed on the label on the container.

Eating, drinking and smoking in areas where treatment work is in progress is not advisable. Any part of the body exposed to woodworm fluids should be washed as soon as possible and contaminated eyes flushed immediately with cold running water.

Care in the use of fluids

Extra care should be taken in kitchens and larders to ensure that the fluids used do not come in contact with exposed food, table-tops, cooking utensils and food containers. Throughout the house, keep fluids away from domestic animals, fish and plants.

Do not dispose of waste woodworm fluid in drains or ditches, or on land adjacent to rivers, streams and ponds.

11 Guarantees and insurance

Although it is unfortunate that it is not possible to observe a woodworm attack until the end of the first life-cycle, when the first adults emerge from the wood through the flight holes they have made, infestations of wood-boring beetle larvae do show certain advantageous features as far as their control is concerned when compared with other forms of insect control. Perhaps the most important concerns the fact that during almost all the life-cycle of wood-boring beetles they are virtually imprisoned in the wood. This simplifies, and indeed makes certain, their complete control. This is because it is possible to apply insecticides to wooden surfaces with great precision as regards siting and lethal dose. Even when larvae are boring deep in the wooden structure, so that they are beyond the effect of the penetrating fluid, they must inevitably return towards the surface in order to construct the pupal chamber and finally emerge as adult beetles. If, therefore, the timber surface has been treated at an accurate dosage rate with a carefully calculated lethal zone of woodworm-killing insecticide the wood-boring larva must inevitably succumb.

Another important advantage possessed by woodworm-killing fluids of the oil-soluble type really concerns the properties of wood itself. This is the ability of wood, and more especially the sapwood, to absorb oil solutions. It is sometimes possible to demonstrate quite remarkable penetration powers of such solutions in sapwood. To a somewhat lesser extent this is true of water solutions also, but in this case the wetter the wood the better the penetrating power of the water solution. This pheno-menon has two effects, the first of which is the obvious one of allowing the woodworm-killing fluid to reach a high proportion of the burrowing larvae. The second concerns the lasting properties of several insecticidal substances. Certain of such substances lose their insecticidal properties after a few months when applied as a superficial film on a non-absorbent base, but when absorbed into wood through the medium of penetrating oil solvents, their woodworm-killing properties last for many years, enabling woodworm servicing companies to offer a guarantee of 20–30 years.

Rentokil

Woodworm & Dry Rot Service

RENTOKIL LIMITED
FELCOURT, EAST GRINSTEAD, WEST SUSSEX RH19 2JY

Certificate of Guarantee

Client...

In any future communication please quote Contract No. ....

Property...

Date of Survey Report.................................. Date of Completion of Treatment...............................

1. Pursuant to the terms of the above-mentioned contract this Guarantee is issued by **RENTOKIL LIMITED**, hereinafter referred to as **Rentokil**, in respect of the work carried out in the above named property by **Rentokil's** technicians in accordance with the recommendations of **Rentokil's** technical advisers as detailed in the Survey Report, Specification and Estimate.

2. **Rentokil** hereby GUARANTEES for a period of 30 years from the date of completion of the treatment in the above property that in the event of any continuation or recurrence of infestation by wood-boring insects or attack by wood-rotting fungi *in any of the timbers treated against such insects or fungi respectively by Rentokil's technicians*, **Rentokil** will upon production of this Certificate, the Survey Report and the Estimate, at a mutually convenient time and without charge, expose and inspect such timbers, determine the then current extent(s) of that continued or recurrent infestation or attack and then execute such further treatment and reinstatement work as is necessary to eradicate (i) that infestation by wood-boring insects in those timbers and/or (ii) that attack by wood-rotting fungi in and beyond those timbers.

3. The client shall retain any pre-existing rights at common law or by statute notwithstanding the terms of this Certificate of Guarantee, which shall be regarded as in addition to and not in diminution of such rights.

4. In the event of any transfer of ownership of the property covered by this Guarantee, **Rentokil** is prepared on application, to recognise the successor in title as the person entitled to the benefit of this Guarantee for its remaining term.

For and on behalf of –
RENTOKIL LIMITED

Date ....

N.B. The client's attention is drawn to the fact that (except as provided in 2(ii) above) the benefit of this Guarantee is limited to the timbers which were the subject of the contract works. The client should consider indemnifying himself against the costs of treating any other infestations or attacks, which might subsequently be suffered by *untreated timbers*, by means of an insurance policy such as can be obtained from **Rentokil Insurance Limited** for many types of property.

This is a valuable document. Please retain it with the other contract documents in a place of safe-keeping.

Copyright **Rentokil Limited** 1980 ©   Ref. W2a 1/80

Fig. 57. The certificate of guarantee supplied after a woodworm treatment by Rentokil Ltd.

The reader may wonder how it is known that treatment will last so long. The answer lies in a test method which enables scientists to accelerate the ageing process in treated blocks of wood. The samples are kept in a wind tunnel at constant air speed, temperature and humidity: nine months in these conditions are found to equal three years in a normal roof void. Further tests with *Anobium* larvae plus some calculation show that sufficient insecticide will remain in the timber after 30 years to kill any invading woodworm larvae.

Thus it is that the long-term guarantee concerning suitable treated wood is not only a practical proposition based on well-known scientific fact but has come to be generally expected as part of a woodworm control service.

The form of the guarantee
It is obvious that such a guarantee must be given only in respect of the treated timbers, which is the reason why a report detailing the timber to be treated is an integral part of the woodworm-control contract. In the author's experience it is a very rare event for re-treatment of timber still showing emergence of wood-boring beetles to be necessary.

In the certificate of guarantee (Fig. 57) issued by the originators of long-term guarantees there is an important clause of particular interest to those selling their home; this is to the effect that in this event, the servicing company is prepared to assign the benefit of the remaining period of the guarantee to the purchaser of the property and will recognize the purchaser as the person entitled to the benefit of the guarantee. The guarantee is additional to the client's rights at common law.

Woodworm and the law
The occurrence of heavy and widespread damage to the structural timbers of buildings and to furniture and the subsequent fear of such damage by the public often leads to disputes. There are the landlord and tenant disputes – the allegation that woodworm in the tenant's furniture has spread to landlord's floors, or *vice versa*. The only comment that can be made is that such spread is impossible to prove with certainty, however *probable* it might appear in some circumstances. To guard against such eventuality if one has either infested premises to lease or infested furniture to put in them, it is best to follow the advice given with regard to application of suitable woodworm-killing fluids.

The selling of premises and furniture infested with woodworm gives rise to another set of problems which may end in litigation. It is not proposed to go into a great deal of detail in this regard, but some sound advice emerges from the writer's experience as an expert witness.

Never guarantee or otherwise give a warranty that any article of wood, be it furniture, structural timber or musical instrument, is free from woodworm *merely because flight holes are absent*. The reader will no doubt by now have realized that it is possible for an article of wood to have a number of woodworm larvae tunnelling in it, yet flight holes would not appear until such larvae had completed their development and emerged as beetles. This has already been decided in the courts against the seller of a piano giving such a warranty.

If you are selling a house or a valuable piece of furniture which you *know* to have woodworm do not try to cover the flight holes or otherwise mislead the intending purchaser in any way concerning its occurrence. You may, you think, complete the sale satisfactorily, but the purchaser may be one of the growing public becoming aware of woodworm, and perhaps some little time after the completion of the transaction the purchaser may be able to prove that you, as the vendor, made out that the house or furniture was other than what it was. Judgements have already been given on such cases which, although rather complicated, indicate clearly that you, as vendor, would not fare too happily in Court.

Before offering such a house or furniture for sale you would be well advised to consult a specialist firm able to survey and estimate for the extermination of the woodworm. Armed with this figure you would then allow such an amount to the intending purchaser, to be deducted from the purchase price.

If you yourself carry out any treatment in your house against wood-worm, or if a specialist firm carries out such work on your behalf, it is important to *keep all the documents*. Attach the receipted bills for wood-worm-killing fluids, etc., to the documents of your house. At the very least they may be well worth their face value in the event of you or your heirs selling the property at a later date.

Guaranteed treatment by a servicing company

Having arrived thus far in the book, the reader will possess sufficient knowledge of the biology of wood-boring beetles to appreciate two important and fundamental facts regarding woodworm infestation by the Common furniture beetle, whether it concerns a piece of furniture or the structural timbers of a building.

First, flight holes merely show where the previous generations of beet-les emerged. Secondly, flight holes do not indicate the total distribution of the present generation of larvae. These are boring away out of sight in the wood, and will probably continue to do so for another year or two before they, in turn, complete their metamorphosis and bite their way out of the timber to the outside world as adult beetles.

When it is remembered that the female lays from twenty to sixty eggs, sometimes over a wide area, it does not take much imagination to realize that the unseen distribution is usually far greater and probably more widespread than the original flight holes suggested. This fact alone immediately makes every woodworm attack, particularly where it is in structural timbers of a property, a two-fold problem. The first part of the problem concerns those timbers which show evidence of attack; i.e. the presence of flight holes. These timbers, and usually those in the vicinity of the attack, making up a separate building unit, i.e. the roof void, will be treated insecticidally.

Provided this treatment is carried out by a reputable firm specializing in this form of remedial treatment and provided also that the treatment is covered by an adequate thirty-year guarantee, then the property owner has solved the first part of the problem, and has nothing further to worry about, at least as far as the treated timbers are concerned. The virtual certainty of woodworm extermination, when carried out in this way has been commented on above.

We must now consider the second part of the problem, the remaining parts of the house not treated insecticidally but where larvae may be present although not visible as yet. To make this clear, let us imagine for a moment that the reader has just bought a house. It is about twenty years old, with an accessible roof void, four bedrooms, bathroom and the usual number of reception rooms on the ground floor.

As a result of the surveyor's inspection, a fairly heavy and widespread attack by Common furniture beetle was discovered in some of the rafters, joists and the boarding of the roof. One of the bedroom floors was also attacked, but not quite to the same extent. These timbers showed evidence of flight holes and were therefore known to be attacked, and on the recommendation of the surveyor, treatment was carried out to the roof timbers and the bedroom floor by a firm giving a thirty-year guarantee.

If we stop now for a moment and analyse the report, which stated that the attack in the roof timbers was fairly heavy and widespread, it must be obvious, first, that the attack is of more than one generation. Secondly, the infestation in the bedroom floor was probably a direct result of the attack spreading from the roof timbers. Remember, the beetles are free-flying insects, and the floor of the bedroom became infested due to the attack in the roof timbers. Once these two facts are put together we can then ask: how much further has the attack spread? What about the floors of the other rooms? This is the question, of course, that brings us to the second part of the problem; those timbers which appear to be free of infestation because there are no flight holes to be seen but which may in fact already be infested by the early larval stage. Even if they were not, these untreated timbers are susceptible to attack coming from elsewhere, and the possibility is that they may become infested at some time in the future. What can we do about these untreated and susceptible timbers? How can we safeguard them and also the pockets of the householder from possible further heavy financial outlay for remedial treatment in the future?

Woodworm insurance

Fortunately, there is a solution to this second part of the problem, and what is probably more important, an answer both simple and inexpensive. It is to insure against it. To be able to insure one's house against

Fig. 58. A proposal form for insurance against woodworm and woodrot.

woodworm damage is a relatively new idea in this country, but in Scandinavian countries it has been possible to do so for a number of years; in fact, in Denmark, woodworm insurance has been in force for more than fifty years. In Denmark, however, the House longhorn beetle is the hazard. In Britain this form of insurance is effected by completing a proposal form (Fig. 58) and returning it to the Company. The Insurance Company will then arrange to carry out a free preliminary survey of the timbers of the property. If no evidence of infestation is revealed then the policy is written without further delay. On the other hand, if there is evidence of an attack, then the infested areas must be treated and paid in full before the insurance can be granted.

One might ask, what are the benefits to be derived from this form of insurance? First, the owner of the property is relieved from worry and future financial outlay, apart from the fact that the capital invested in the property is protected at reasonable premium rates. Many householders would also regard the free preliminary inspection as a benefit, and the fact that the property is insured against attack must surely be an added selling point, should the owner decide to sell his property. A woodworm insurance policy is also assignable by notifying the Insurance Company of the change of ownership.

12 Termites

If the whole world is considered, then without a doubt by far the most important wood-destroying animals are termites. About 1800 different species occur in countless numbers in all the warm countries of the world. Happily, in this instance, neither Great Britain nor indeed Northern Europe comes into this category. Two species, however, *Kalotermes flavicollis* and *Reticulitermes lucifugus*, are found commonly in Southern Europe, and another species infests a small area of the city of Hamburg. It is in view of the fact that all articles, constructed wholly or in part of wood, intended for use in tropical or sub-tropical countries, must be rendered termite-proof that this short account is given here. In addition to wood, many other materials including certain plastic products, are often damaged by termites.

Termites are often given the name of 'White Ants'. Actually they are not ants and, in fact, are very dissimilar in structure and thus, classification. It is, however, very probable that in the habits of the termites living in very large communities the layman has seen a similarity to the social organization of the true Ants (FORMICIDAE).

The order of insects to which the Termites belong is known as the ISOPTERA. They do not undergo a metamorphosis during development, as do beetles and butterflies, where the grub or larval stage is very unlike the adult insect. The young termites when they hatch from eggs are very like the adult in form, and any differences that do occur are assumed more or less gradually during the development period, as they shed their successive skins.

Termites are very interesting biologically on account of the complex social organization in which they live. In the case of what are usually known as subterranean termites, the adults exist in a number (usually three or more) of what are termed 'castes'. These are the forms in which they exist. King and Queen termites are usually produced from winged forms. These forms possess two pairs of almost identical wings which are of service for a few minutes only during the sexual swarming. When, however, they drop to the ground and pair, the wings break off close to the base (Fig. 59), as thereafter they live in a special chamber deep in the colony, from which they do not emerge again. The abdomen of the

Fig. 59. Queen termite. Note stumps of wings where they have broken off.

Queen enlarges to immense proportions – she produces many tens of thousands of eggs. From her eggs are produced worker and soldier termites which never attain sexual maturity. These are produced in very large numbers and spend their whole lives tending the communities in which they live. The workers forage for food, enlarge the nest and keep it clean (which involves eating the bodies of their own kind), or in the case of the wood-boring species carry out all tunnelling. The wood they eat is digested by PROTOZOA (one-celled animals) in the stomach and a salivary material produced from the digested wood is fed to the King and Queen, whose digestive organs are modified to take only this food.

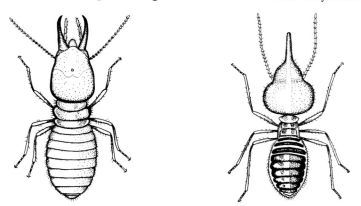

Fig. 60. Soldier termites: (left) mandibulate type; (right) nasute type. They defend the colony but do no other work.

The sole duty of the soldier termite is the defence of the colony (Fig. 60). On being disturbed the soldiers take up threatening attitudes in the front line and will seize in their mandibles any marauding insects such as true ants whilst the workers immediately stow young brood and eggs in the deeper recesses of the colony. The soldiers of some termite species are able to eject a stream of fluid which repels the attackers. These are known as nasutes (Fig. 60) and they possess a large beak-like projection at the front of the head which secretes a strong-smelling and sticky fluid from a gland at the tip.

From time to time another caste – subsidiary reproductives – are produced; these are to augment egg production. They differ from the Queen in never having possessed wings. When climatic conditions are propitious the colony swarms. Large quantities of 'alate' or winged reproductives are produced which, after a brief spell of flying, mate, and then shed their wings before searching out a crevice in a piece of wood in order to start a new colony.

Termite classification, with notes on main genera
The termites are classified into six families. The MASTOTERMITIDAE contains only a single Australian species found in tree stumps underground. The community often consists of over a million individuals, but there is doubt as to whether these are true workers. Fossil forms of species in this family have been found in England, and the family is important as showing the relationship between the termites and the cockroaches.

Members of the second family, the KALOTERMITIDAE, are known as the drywood termites, on account of their inhabiting dry wood which may have no direct connection with the ground. Usually the communities are small, and there is no worker caste, the individuals mostly being juveniles (Fig. 61). Soldiers, however, are present which are characterized by the

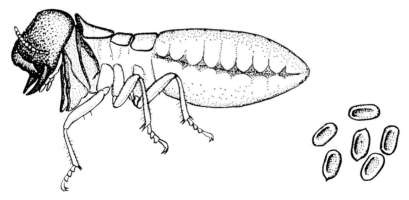

Fig. 61. A drywood termite *Cryptotermes brevis* (length: 4·5 mm) with characteristic faecal pellets.

possession of large heads and strong mandibles. The faecal pellets are also characteristic, being dry and seed-like, and from time to time a quantity is ejected from the wood through a hole made for the purpose which is usually closed up again afterwards. Well-known species are *Kalotermes flavicollis* from Southern Europe, *Cryptotermes brevis* the West Indian drywood termite, and *Cryptotermes dudleyi*, from East Africa, Central America and South-East Asia.

The TERMOPSIDAE are known as the dampwood termites and are usually found in fallen logs and stumps in conditions of high humidity. Like the KALOTERMITIDAE, the colony size is small and true workers are not present. One species, *Zootermopsis angusticollis*, gets transported over long distances, sometimes reaching this country in logs of Douglas Fir and Californian Redwood from the Pacific coast of North America.

The HODOTERMITIDAE, the fourth family, are called the Harvester termites.

The great majority of termite species are classified in the family TERMITIDAE. In the sub-family MACROTERMITINAE wood is chewed into 'combs' on which certain species of fungi are grown and devoured. The *Microcerotermes* are found throughout the tropics, and *Amitermes* are particularly able to live in desert regions. The *Odontotermes* are of considerable economic importance. The *Nasutitermes* are remarkable for the pear-shaped head of the soldiers, and some species build arboreal nests.

The last family, the RHINOTERMITIDAE, are the subterranean termites. They are usually small and are important wood destroyers; *Reticulitermes lucifugus* is the destructive European species, and *flavipes* in the same genus destroys much timber in buildings in the U.S.A.

The main colony or nest of this type of termite is located underground, often very deep, or in otherwise inaccessible places. From such a situation the workers make long galleries through the earth until some wood is located when the tunnel is continued into the timber. Only rarely is evidence given of the presence of the termites until the wood is so badly eaten away by the tunnelling that it collapses or breaks. Termites never work in the light, nor, indeed, in the open; they require a very close control of humidity, and if any open space has to be crossed it is accomplished by the construction of a tunnel or tube, which can be seen, made of earth and saliva. In this way the subterranean termites can bridge over concrete building foundations and commence an attack on sound woodwork on the ground floor.

Control

The termite problem in the warm countries is such a large one that man's effort to safeguard timber in all its uses has resulted in a vast quantity of literature involving all types of recommendations. It is, however, amazing what little concern is shown with regard to this problem in many areas. Timber and labour is cheap and infested woodwork is completely replaced rather than simple preservative treatment being carried out.

For the reader in Britain we give below brief precautions that can be taken in the termite-infested territories, but what will be more the concern of the reader will be the precautions to take regarding timber-fabricated articles – including packing cases – manufactured in Britain but destined for the tropical and sub-tropical countries.

In areas where termites abound no timber in whatever form should be allowed to remain in contact with the ground. Keep the ground clear of all rubbish, fallen twigs, etc. Before building commences a soil pre-treatment operation should be undertaken. This consists of saturating thoroughly with long-lasting modern insecticides the soil in the area where the building is to be constructed. In the construction of houses and other buildings allow no timber to be used within 50 cm of the ground. Use bricks or concrete for foundations. Constant vigilance must be maintained in order to see that no shelter-tubes are constructed over the foundations.

All timber used in building construction should have two or three applications of an oil-soluble wood preservative manufactured for the purpose or should be pressure impregnated with a water-borne preservative. There are a number of proprietary brands. Dipping with oil-based preservatives is often carried out and can also be recommended.

For the British manufacturer producing articles wholly or partly of wood for use in tropical or sub-tropical countries the specification should include complete treatment of wooden parts with oil-soluble or water-borne wood preservatives. All wooden packing cases should be similarly treated.

Treatment should be carried out when the wooden article is completely fabricated; that is, when no fresh wood surfaces will be produced. Furniture destined for abroad, as well as motor truck chassis and scientific instruments, should be treated in this manner.

13 Insects commonly confused with woodworm

Almost every insect which has at one time or another made its way into a house accidentally or otherwise must have been confused with woodworm – at least, that is what the writer believes after receiving specimens for identification from many tens of thousands of householders.

These insects fall into two main categories – those which are infesting some materials in the home (other than wood) such as textiles or stored food, and another group consisting of wanderers from the garden or elsewhere – their presence in the house being accidental or fortuitous. The following brief description includes insects which, from experience, the writer knows to be most commonly but erroneously thought to be wood-boring pests. No attempt is made here to produce an entomological textbook, but the illustrations and short notes should make identification of the household insects quite simple in most cases.

Biscuit beetle *Stegobium paniceum*
This beetle is in the same family – ANOBIIDAE – as the Common furniture beetle and Death watch and it does have a strong resemblance to the former (Fig. 62), and, indeed, it is often mistaken for it by householders. It is not a wood-borer but is a common pest of many stored foods, especially farinaceous ones. In warm kitchens or cupboards it often emerges at odd times during the year. It sometimes subsists on crumbs which have collected between the floorboards of old houses.

It can be differentiated from the Common furniture beetle by its more rounded appearance. It is always a little more reddish and is covered with short dense fair hair when freshly emerged. The lines of punctures along the wing-cases are not so deep. Householders often find the beetles, recognize their general similarity to pictures of Common furniture beetle, but find no flight holes. In such cases it is not difficult to track down the material which is infested, as often a few of the beetles are found hiding in the folds of the paper wrapping. It is a good plan to destroy the infested food, or include it in a chicken mash and then spray the inside of the cupboard and surroundings with an odourless oil-solu-

Fig. 62. The Biscuit beetle, *Stegobium paniceum*, a near relative of the Common furniture beetle, but is not a wood-borer. Length: 2·5 mm.

ble insecticide. Do not use woodworm-killing fluids where food is stored.

Plaster and Fungus beetles
A relatively large number of beetles have larvae which feed on fungal hyphae, so that where damp conditions favourable to the growth of fungi prevail, some species of these beetles may be attracted. They are sometimes present in enormous numbers in new houses or where new plastering has taken place. The damp plaster causes moulds or mildews to appear on paper, woodwork or textile fibres, and it is on these fungi that the adult beetle lays her eggs and on which the larvae subsequently feed. They are found also feeding on moulds growing on imperfectly-ventilated stored foodstuffs, such as jam or cheese, and in damp basements. The common name of Plaster beetle is thus a misnomer.

Three species are very common, namely *Lathridius minutus*, *Aridius nodifer* and *Cryptophagus acutangulus*. All these species (particularly the first-named) are found on woodwork and today are common timber-yard insects. Indeed, many timber-yard men confuse it with *Lyctus*, although *Lathridius minutus* is only a fraction of the size of *Lyctus*. Plaster beetles do not damage woodwork.

Control (indoors)   Dry out new plaster as soon as possible by keeping the rooms warm and well-ventilated. Wash newly plastered walls down with a fungicide, such as a solution of sodium pentachlorphenate in water, if it would appear that the moisture in the wall would dry out slowly.

With regard to timber (fresh sawn, 'wet' obeche seems very susceptible to *Lathridius minutus*) arrangements should be made to dry it out as
quickly as possible.

Spider beetles

A number of beetles in the family PTINIDAE, owing to their superficial
resemblance to spiders, are known as Spider beetles.

The species most commonly met with are:

Ptinus tectus	Australian spider beetle (Fig. 63)
Niptus hololeucus	Golden spider beetle
Ptinus fur	White-marked spider beetle

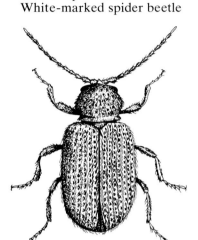

Fig. 63. The Australian spider beetle (*Ptinus tectus*), originally a native of
Tasmania, now spread all over the world as a pest of stored foods. Length:
3–4 mm.

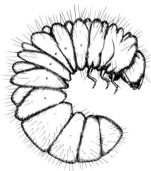

Fig. 64. The grub of the Australian spider beetle. Known to British soldiers during the First World War as the inhabitant and sharer of their
Army biscuits.

They are all pests of stored products, but are often found in the older type of house, where the larvae subsist on bread-crumbs and similar farinaceous material lodged in crevices between the floorboards. They are however able to thrive on a wide variety of materials, including mouse droppings, and probably wool dust is also eaten by them. It is difficult to account for the occasional occurrence of swarms of the Golden spider beetle. The larvae of all the Spider beetles (Fig. 64) are creamish in colour and roll themselves into a ball if disturbed.

Generally, it can be said that Spider beetles do not damage wood, but occasionally the larvae do hollow out pupal chambers in wooden surfaces.

Control Find the source of infestation and destroy, if possible, any food material containing the live stages. Spray regularly with oil-based insecticides all crevices in woodwork, etc., in the vicinity of the infestation.

Silverfish *Lepisma saccharina*
The Silverfish insect is about 12 mm long, cylindrical, rather torpedo-shaped and with three long bristles emerging from the tail-end (Fig. 65). It is metallic grey in colour, perhaps more like lead than silver. It is very agile, seeking cover extremely rapidly when disturbed in its usual dark quarters. It is especially fond of warm damp places. It does not damage woodwork; indeed, it is seldom that any damaged material is observed which can be put down to Silverfish. It is closely related to an insect known as the Fire brat, *Thermobia domestica*, often quite common in factories, bakehouses, etc., where steam heating is employed.

Fig. 65. Silverfish. Metallic grey in colour. Length: 12 mm. Does no damage to woodwork and little damage at all.

Control Liberally dust with powder insecticide and use powder blower to distribute the powder in the crevices in brickwork and plaster used by the insects.

Hide beetles DERMESTIDAE

These beetles are notable pests of stored food products, skins, furs and materials of high protein content. A common species is *Dermestes lardarius*, known as the Larder beetle or Bacon beetle. The larva has a banded appearance and is beset with hairs of various lengths. The larva should never be handled, as some of the hairs break off in the skin and set up considerable irritation. Just before pupation takes place the larva burrows into any convenient compact material, and wood is frequently chosen. The diameter of the ingress hole and tunnel is about 3 mm, circular, and may be anything from about 25 mm to about 30 cm in length. Hide beetle galleries in wood are, however, only common in such places where the larvae are infesting dried skins or dried meat products such as in warehouses and tanyards. The banded cast larval skins are often present in the holes.

The adult beetles vary from seven to nine millimetres in length and are easily identified on account of the characteristic pattern of greyish-gold and black shown in the accompanying illustration (Fig. 66).

Control Thoroughly clean out premises where Larder beetles occur and burn débris and infested material. Treat all exposed woodwork with

Fig. 66. The Larder beetle, *Dermestes lardarius*, sometimes a wood-borer – when searching for a site for pupation. Length: 6–10 mm.

a woodworm-killing fluid. Spray all materials susceptible to infestation regularly for a period – say, once a week for two months – with an oil-soluble insecticide.

Carpet beetles

The larvae of Carpet beetles (belonging to the beetle family DERMESTI-DAE) are important pests of woollen material and furnishings in most parts of the world. Carpets are the usual furnishings attacked; that is why this group of insect pests is so named.

Fig. 67. Side view of a grub of *Attagenus pellio*. Length: 10 mm.

Grubs of the *Attagenus* group are rather long and slender, greyish-brown or light-brown in colour, with a tuft of long hairs on the tail region (Fig. 67). They are often found underneath the edges of carpets (especially fitted carpets) and the banded appearance of the cast or moulted skins is quite characteristic.

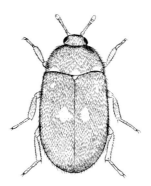

Fig. 68. The Fur beetle. *Attagenus pellio*. Length: 4–6 mm.

The adult beetle of *Attagenus pellio* is usually referred to as the Fur beetle (Fig. 68). It is about 5 mm long, and black in colour except for three indistinct greyish marks on the prothorax and two distinct grey spots on the wing-cases (almost in the middle of the back).

On the other hand, the grubs of the *Anthrenus* group are much fatter in appearance, with long body hairs and several tufts of long hairs in the

Fig. 69. The 'woolly bear' grub of *Anthrenus verbasci*. Length: 5 mm.

tail region (Fig. 69). They are often known as 'woolly bears'. Their hairy appearance blends with the carpets or materials which they are attacking. In addition, they often hide in cracks in wooden floorings and clothes cupboards, etc. In airing cupboards they seem very resistant to dryness, and are able to withstand such dry conditions which clothes moth grubs could never tolerate. The adult beetles into which the grubs develop are rather globular, somewhat like the Ladybird beetles (*Coccinella*). The various species, however, have different patterns of black,

Fig. 70. The Varied carpet beetle. Length: 2·5 mm.

white and grey markings. The species illustrated (Fig. 70), is *Anthrenus verbasci*, usually called the Varied Carpet beetle.

The grubs of Carpet beetles are very much more difficult to eradicate with complete success than the Clothes moth, due partly to their habit of secreting themselves in tiny cracks in woodwork and partly to the scaly effect of their bristles, which seem to hold off powder insecticides and many sprays. But, if regular spraying of all woollen materials is carried out with a mothproofing fluid (specially formulated for moth control), eradication will be effected. Care should be taken, however, to spray over all – not to give a soaking in one place and nothing in another.

Furniture mite and Booklice

A few years ago Furniture mite was a serious inconvenience to many householders. It is indeed strange to have to record that the Furniture mite does not appear to be so common today as it was formerly. On the other hand, the Booklouse is today exceedingly common, and throughout the summer months innumerable requests are received for the identification and treatment in order to eradicate 'small creamy-white, quick-running insects about 1 mm long which run over the surface of the furniture and then hide in cracks and crevices'.

From this description it is easy to determine the cause of the annoyance as Booklice or *Psocids* (Fig. 71). They have no connection, of

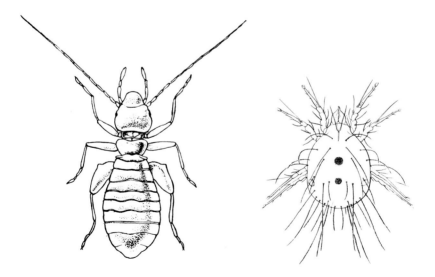

Fig. 71. (Left) a Booklouse or Psocid, 1 mm long; (right) a Furniture mite, 0·5 mm long. Both feed on minute fungi on damp surfaces.

course, with wood-boring insects such as the Common furniture beetle, although they are often found in association. They cannot be said to cause any damage to furnishings. It is believed that they feed only on minute fungi, which, invisible to the naked eye, occur on many types of surface. It is a cause of considerable annoyance to the housewife to see those white insects running about on the furniture. In fact, the reaction of the average person on seeing them for the first time is that the insects are something of a much more destructive nature.

The life-cycle of the Booklouse is very simple. Eggs are laid by the females amongst dust in crevices in woodwork and after a short time the young Psocid hatches. It does not, like many insects, undergo a larval or grub stage but is in most respects very similar to the adult (except for size and sexual maturity). As it feeds, it changes its skin several times, and during the summer months takes only a few weeks to reach the adult stage.

To eradicate Psocids is fairly easy. They are usually indicative of warm, humid conditions prevailing, and all windows in a room where they occur should be kept open for several days, day and night for preference. Thereafter, treatment with the fluid normally used for wood-worm treatment is sufficient to destroy them completely. The fluid should be applied by brush, special attention being paid to joints and cracks in the woodwork where the Booklice will be seen to scurry out and succumb within a few seconds or so. Tests carried out by the writer on Psocid destruction in antique furniture gave remarkable results which were both gratifying and extremely successful.

Conditions which favour the increase of Booklice also favour the

Fig. 72. The Woodlouse, an isopod crustacean. About 15 mm long. It feeds on vegetable debris, usually much decayed.

increase of the Furniture mites, and perfectly dry furnishings (especially upholstered suites) brought into damp newly-built houses often bring about an attack by Furniture mite. These mites are only just visible to the naked eye and appear as minute white specks (Fig. 71) when seen on dark polished wood. They move very slowly, so that it is easy to differentiate between this pest (not an insect; it belongs to the spider group) and the much larger, quick-running Booklouse.

Woodlice   *Porcellio scaber* and *Oniscus asellus*
The Woodlouse is not an insect but a crustacean (in common with shrimps and crabs). It is characterized by its ten tile-like segments which articulate with each other and, on the underside, its seven pairs of walking legs. From the head arise a pair of large jointed antennae (Fig. 72). It is rather variable in colour, from bluish-grey to greyish-brown. The species *Armadillidium vulgare* can roll itself into a complete ball when disturbed. All species feed on vegetable débris of various sorts and frequent damp and dark places; in fact, very humid conditions are essential for its existence. This accounts for it being found in damp places in kitchens, near drains, and sometimes actually in the bath. It causes no damage to sound wood but is often found where wood is in the last stages of fungal decay. It is a nuisance but completely harmless in the home.

Bibliography

Below is given a selection of books and leaflets in which the reader will find various aspects of the woodworm problem dealt with in greater detail than is possible in the present book.

B.R.E., Princes Risborough Laboratory technical publications include:
Technical Note No. 39 (1973) *The house longhorn beetle.*
Technical Note No. 45 (1974) *The death watch beetle.*
Technical Note No. 47 (1974) *The common furniture beetle.*
Technical Note No. 55 (1972) *Defects caused by ambrosia (pin-hole borer) beetles – origin and recognition.*
Technical Note No. 60 (1975) *Lyctus powder-post beetles.*

COGGINS, C. R. 1980. *Decay of timber in buildings.* Rentokil Ltd. (A companion volume to the present book which deals with dry rot, wet rot and other fungi.)

DUFFY, E. A. J. 1953. *A monograph of the immature stages of the British and imported timber beetles (Cerambycidae).* Brit. Mus. (N.H.), London. (For the specialist.)

HARRIS, W. V. 1971. Termites. *Their recognition and control.* 2nd edn. Longmans, London.

HICKIN, N. E. 1967 *The conservation of building timbers.* Hutchinson, London.

HICKIN, N. E. 1974. *Household insect pests.* 2nd edn. Associated Business Programmes, London.

HICKIN, N. E. 1975. *The insect factor in wood decay.* 3rd edn. Associated Business Programmes, London. (The definitive work on wood-boring insects.)

HICKIN, N. E. 1971. *Termites – a world problem.* Hutchinson, London.

HICKIN, N. E. 1971. *Wood preservation – A guide to the meaning of terms.* Hutchinson, London.

WILKINSON, J. G. 1979. *Industrial timber preservation.* Associated Business Press, London. (A technical guide to the treatment of timber prior to use.)

Subject Index